Human Anatomy and Physiology Coloring Workbook and Study Guide

The Jones and Bartlett Series in Biology

Basic Genetics
Daniel L. Hartl, Washington University School of Medicine; David Freifelder, University of California, San Diego; Leon A. Snyder, University of Minnesota, St. Paul

The Biology of AIDS
Hung Fan, Ross F. Conner, and Luis P. Villarreal, all of the University of California, Irvine

Cancer: A Biological and Clinical Introduction,
Second Edition
Steven B. Oppenheimer, California State University, Northridge

Cells: Principles of Molecular Structure and Function
David M. Prescott, University of Colorado, Boulder

Cross Cultural Perspectives in Medical Ethics: Readings
Robert M. Veatch, editor, The Kennedy Institute of Ethics - Georgetown University

Early Life
Lynn Margulis, University of Massachusetts

The Environment, Third Edition
Penelope Revelle, Essex Community College; Charles Revelle, The Johns Hopkins University

Essentials of Molecular Biology
David Freifelder, University of California, San Diego

Evolution
Monroe W. Strickberger, University of Missouri, St. Louis

Experimental Techniques in Bacterial Genetics
Stanley R. Maloy, University of Illinois, Urbana

Functional Diversity of Plants in the Sea and on Land
A. R. O. Chapman, Dalhousie University

General Genetics
Leon A. Snyder, University of Minnesota, St. Paul; David Freifelder, University of California, San Diego; Daniel L. Hartl, Washington University

Genetics
John R. S. Fincham, University of Edinburgh

Genetics of Populations
Philip W. Hedrick, University of Kansas

Handbook of Protoctista
Lynn Margulis, John O. Corliss, Michael Melkonian, and David J. Chapman, editors

Human Anatomy and Physiology Coloring Workbook and Study Guide
Paul D. Anderson, Massachusetts Bay Community College

Human Genetics: A New Synthesis
Gordon Edlin, University of California, Davis

Introduction to Biology: A Human Perspective
Donald J. Farish, California State University at Sonoma

Introduction to Human Disease, Second Edition
Leonard V. Crowley, M.D., St. Mary's Hospital, Minn.

Introduction to Human Immunology
Teresa L. Huffer, Shady Grove Adventist Hospital, Gaithersburg, Maryland, and Frederick Community College, Frederick, Maryland; Dorothy J. Kanapa, National Cancer Institute, Frederick, Maryland; George W. Stevenson, Northwestern University Medical Center, Chicago

Living Images
Gene Shih and Richard Kessel

100 Years Exploring Life, 1888-1988, The Marine Biological Laboratory at Woods Hole
Jane Maienschein, Arizona State University

Medical Biochemistry
N. V. Bhagavan, John A. Burns School of Medicine, University of Hawaii at Manoa

Medical Ethics
Robert M. Veatch, editor, The Kennedy Institute of Ethics - Georgetown University

Methods for Cloning and Analysis of Eukaryotic Genes
Al Bothwell, Yale University School of Medicine; George D. Yancopoulos, Regeneron Pharmaceuticals; Frederick W. Alt, Columbia University, School of Physicians and Surgeons

Microbial Genetics
David Freifelder, University of California, San Diego

Molecular Biology, Second Edition
David Freifelder, University of California, San Diego

The Molecular Biology of Bacterial Growth
(a symposium volume)
M. Schaechter, Tufts University Medical School; F. Neidhardt, University of Michigan; J. Ingraham, University of California, Davis; N. O. Kjeldgaard, University of Aarhus, Denmark, editors

Molecular Evolution: An Annotated Reader
Eric Terzaghi, Adam S. Wilkins, and David Penny, all of Massey University, New Zealand

Oncogenes
Geoffrey M. Cooper, Dana-Farber Cancer Institute, Harvard Medical School

Plant Nutrition: An Introduction to Current Concepts
A. D. M. Glass, University of British Columbia

Population Biology
Philip W. Hedrick, University of Kansas

Vertebrates: A Laboratory Text, Second Edition
Norman K. Wessells, Stanford University; Elizabeth M. Center, College of Notre Dame, editors

Virus Structure and Assembly
Sherwood Casjens, University of Utah College of Medicine

Writing a Successful Grant Application, Second Edition
Liane Reif-Lehrer, Tech-Write Consultants/ERIMON Associates

Human Anatomy and Physiology Coloring Workbook and Study Guide

with images from the National Library of Medicine's Visible Human Project

PAUL D. ANDERSON, Ph.D.

Middlesex Community College

With illustrations by

VICTOR M. SPITZER, Ph.D.

University of Colorado School of Medicine
Department of Cellular and Structural Biology
Department of Radiology

JONES AND BARTLETT PUBLISHERS

Sudbury, Massachusetts

BOSTON   LONDON   SINGAPORE

Editorial, Sales, and Customer Service Offices

Jones and Bartlett Publishers
40 Tall Pine Drive
Sudbury, MA 01776
info@jbpub.com
http://www.jbpub.com

Jones and Bartlett Publishers International
Barb House, Barb Mews
London W6 7PA
UK

Portions of this book first appeared in *Human Anatomy and Physiology Coloring Workbook and Study Guide,* ©1990, Jones and Bartlett Publishers, Inc.

Printed in the United States of America

00 99 98 97   10 9 8 7 6 5 4 3 2

ISBN: 0-7637-0499-7

Text illustrations: Kent Leech
Cover design: Marshall Henrichs
Cover illustrations: Victor M. Spitzer

Preface

Human Anatomy and Physiology Coloring Workbook and Study Guide has been written to guide students in the study of introductory anatomy. It is intended to accompany most of the leading anatomy and physiology textbooks. The study guide examines virtually every structure of the human body that is typically studied in an introductory anatomy and physiology course. Because its detailed contents minimize the need for supplemental handouts, the manual is a strong teaching device in itself. The copious, carefully drawn illustrations of the human body systems exhibit structures essential to students' understanding of anatomy.

My goal has been to produce a clear, accurate, and interesting introductory level study guide. Although it is appropriate for anyone curious about the structure of the human body, the book contains features of special interest to those seeking careers in the health sciences. Basic patterns and organizational themes are introduced in the first chapter and are stressed throughout the text. Such concepts are easiest to grasp when presented in an organized framework, and the patterns provide a firm foundation for other courses in the life sciences. Equally important, pattern recognition provides an awareness of the symmetry and logic of scientific thought. Those seeking a career in the sciences will have this concept reinforced in subsequent classes, but for those who pursue other careers, the concept that anatomical and physiological processes are understandable, relevant, and logical should remain intact long after the origins of the latissimus dorsi have been forgotten.

Each chapter is organized into five principal sections:

I. CHAPTER SYNOPSIS. The chapter synopsis provides a concise resume of the contents of each chapter in the text.

II. OBJECTIVES. The objectives section lists goals for each chapter. I have stated them in terms of student achievement rather than instructor performance because educational objectives reflect the aim(s) of a particular course and the value systems of a specific instructor; thus, no one set of objectives will meet all requirements. Nevertheless, you should be able to either select from the lists in Section II those objectives which meet your specific needs or modify them where applicable.

III. IMPORTANT TERMS. A new educational frontier has opened in response to the rapidly growing need for a working knowledge of medical terminology in a variety of medical and paramedical office positions. This current burst of interest is the result of constantly expanding paper work caused by increased population, more national attention to the improvement of health, and ever-increasing medical and hospital insurance coverage.

Since medical terminology as applied to the body's many anatomical systems is a main concern of this book, the use, spelling, English translation, and pronunciation of medical terms are stressed. In the process of studying these four important areas of medical terms, the student will also learn about the structure of each anatomical system and about some of the more common diseases, anomalies, and surgical procedures.

IV. EXERCISES. This study guide offers new opportunities for self-assessment of learning. Regardless of the study situation—alone, with a study partner, after a lecture, after a laboratory, or studying home at night—the student can perform the challenging exercises and apply abstract concepts to reality. Health students can see the relevance of anatomy to current or future clinical experiences.

Some of these exercises ask students to only label a diagram, but most require some coloring of the figure. Soft, colored pencils are recommended so that the underlying diagram shows through. Since most figures have multiple parts to color, students will need to have a variety of colors at their disposal. They may use any color they wish except where specifically directed otherwise, as in the cases of arteries, veins, nerves, and lymphatic vessels, which are normally colored red, blue, yellow, and green, respectively. However, where several different arteries, veins, nerves, or lymphatic vessels are to be colored, other colors must be used (as in the plates of the cardiovascular system). As a general rule, more neutral colors are recommended for smaller or less important areas.

Do not color over the heavier outlines, for they are usually the border lines separating areas to be colored. The lighter lines are usually included to suggest texture or define form in an area to be colored; these lighter lines should be colored over.

V. TEST ITEMS. Section V contains a listing of sample multiple-choice, true-false, and matching questions. These questions have been correlated with the individual chapter contents of most anatomy and physiology textbooks. However, all experienced instructors know the difficulties in constructing valid test questions that determine whether or not the student has acquired mastery of the subject, because valid test questions alone do not assure a valid test. The questions must represent adequate random sampling of the subject area, have a wide range of difficulty, and test fact as well as application of specific facts to everyday situations. Thus, the questions contained herein are simply designed to determine specifically whether the student has mastered the course objectives. The final evaluation of the student is left to the individual instructor.

Once the student has defined the important terms, completed the pertinent exercises, and tried the different tests, a series of puzzles have been prepared to further test mastery of the subject matter. Each chapter will end with a puzzle prepared from the terms at the beginning of the chapter. How do you know if the puzzle has been done correctly? When all the terms fit! LEARNING CAN BE FUN! Enjoy.

I have included a complete glossary of all major human anatomy and physiology terms in the back of the study guide. This feature facilitates independent study in any environment by eliminating the need to have a textbook on hand as a reference.

In closing I would like to thank Dr. Victor M. Spitzer for creating some of the breathtaking three-dimensional images specially for this text. His work represents the cutting edge of the field. A complete text of the National Library of Medicine's images, as generated by Dr. Spitzer: *Atlas of the Visible Human: Reverse Engineering the Human Body*, is also available from Jones and Bartlett Publishers, Inc.

To the Student

How do we learn? From the many theories that have been proposed in answer to this question, most theorists agree on one point: learning requires active participation by the learner. People learn by doing, not merely by listening. Study aids that ask the student to do something with the material he or she is studying are based on the principle of participation.

You probably know from your own experience that the facts and ideas you remember best are those you need to use—not just in the lecture hall or at exam time but in the every day business of your life. To give a simple example, the math skills you use to figure out your budget are remembered after the cotangents and logarithms are forgotten.

In this study guide we offer a means of using the information in the text in various ways, not only to help you organize the material in your mind, but also to help you digest it and make it a part of your way of looking at the world around you. Remember that even so simple an action as taking pencil in hand and writing a fact down in its allotted space will move you one step away from the listening position. Putting that fact into a slightly new relationship with another fact will move you yet another step. The actual process of "digestion" has begun when you have put together all the facts about one subject that are at your disposal and applied them to discover something about a different subject.

Education research has shown that almost 90 percent of the material a student hears or reads is forgotten within a few months—a most discouraging fact! Fortunately, retention can be increased by working with the material and by explaining it to others. This study guide gives you the opportunity to do just that; it requires you to use the material you are studying. To answer many of the questions you must synthesize, analyze, and interpret information. The study guide is designed to help you focus on the important concepts you need to learn anatomy.

You will undoubtedly find your own ways of using this study guide, but here are some suggestions that may be helpful. Before beginning the exercises, read the text chapters carefully, keeping in mind the main points as you read. Then review the synopsis at the beginning of the study guide section. Summarize each concept in your own words. Ask yourself the following questions: Could I explain these points to someone who knew nothing about the subject? With what I know now, could I write a short essay on the major topics of the chapter? Do I really understand the principles involved? In other words, is the overall picture clear in my mind? If you are uncertain of any part, go back and reread that section of the text. Only then should you go on to the exercises.

Check the answers at the end of each chapter when you have finished each part; don't wait until you have completed the whole chapter. In this way, you can correct yourself and build your mastery of the material as you go along. If your answer does not agree with the study guide, go back to the text and find out why. If necessary, ask your instructor for help. You may need clarifications of some points. In any case, before you proceed to the next exercise be sure that you understand the one you have just completed.

One more point about learning. Anyone who is good at sports will agree that the game becomes more interesting as the player's skills improve. A tennis player who can place her serve has more fun on the court than one who keeps lobbing the ball over the back fence. This is equally true in any field of study, including biology. The more you learn about anatomy, the more interesting—even exciting—the subject will become to you.

The author and publisher welcome your comments regarding this study guide and its usefulness to you in your learning. We will especially appreciate any criticisms or suggestions for improvement.

Contents

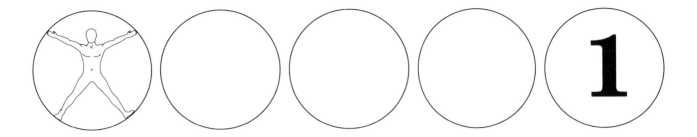

Basic Anatomy

I. CHAPTER SYNOPSIS

This chapter introduces the student to the organizational pattern of the entire body. Among the topics considered are the location and contents of the principal body cavities, the characteristics of the anatomical terms (with common names of body regions), the use of directional terms, planes and sections of the body, and the linear units of measurement in the metric system and how one unit converts to another.

II. OBJECTIVES

After reading this chapter, the student should be able to:

- Describe the pH scale and relate its significance.
- Name the planes of the body using proper anatomical terms.
- Name the important regions of the body and the body cavities.
- Name the various abdominal regions.
- Differentiate between the English and metric systems of measurement by comparing the different units of length.

III. IMPORTANT TERMS

Using your textbook, define the following terms:

abdominal (ab-dom′-in-al) _____

_____

acid (as′-id) _____

_____

acidosis (as-i-do′-sis) _____

_____

alkaline (al′-kah-line) _____

_____

alkalosis (al-kah-lo′-sis) _____

_____

bilateral (bye-lat′-ah-rul) _____

_____

cranial (kray′-nee-al) _____

_____

distal (dis′-tal) _____

_____

dorsal (door′-sal) _____

_____

epigastric (ep-i-gas′-trik) _____

frontal (frunt′-al) _____

_____

hypochondriac (hi-po-kon′-dree-ak) _____

_____

hypogastric (hi-po-gas′-trik) _____

_____

iliac (il′-ee-ak) _____

_____

inferior (in-fear′-e-or) _____

_____

inguinal (ing′-gwin-al) _____

_____

lateral (lat'-ah-ral) _____

_____

lumbar (lum'-bar) _____

_____

medial (meed'-ee-al) _____

_____

midsagittal (mid-saj'-et-al) _____

_____

pelvic (pel'-vik) _____

_____

proximal (prok'-sah-mahl) _____

_____

spinal (spine'-al) _____

_____

thoracic (tho-ras'-ik) _____

_____

transverse (tranz-vers') _____

_____

umbilical (um-bil'-i-kahl) _____

_____

ventral (ven'-trahl) _____

_____

vertebral (vur'-tee-brul) _____

_____

IV. EXERCISES

Complete the following exercises in the order given. A precise set of terms and planes has been chosen to describe positions, relationships, and directions within the human body.

Exercise 1.1

Completion.   Fill in the blanks with the appropriate terms.

Key:
acid
alkaline
H^+
OH^-
neutral

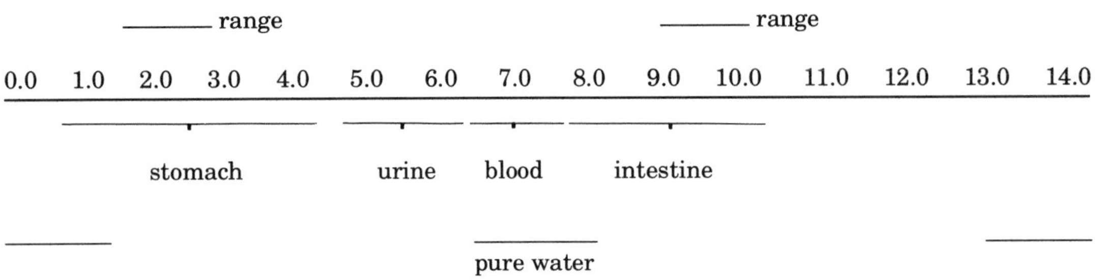

———— range   ———— range

0.0   1.0   2.0   3.0   4.0   5.0   6.0   7.0   8.0   9.0   10.0   11.0   12.0   13.0   14.0

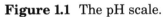

stomach   urine   blood   intestine

————   ————   ————

pure water

Figure 1.1 The pH scale.

Key:

acid

acid-base

acidity

acidosis

alkaline

alkalosis

higher

lower

H^+

OH^-

neutral

pH

It is essential to understand that a number on the _____ scale is actually the result of dividing the numeral 1 by a mathematical value called a logarithm. The result is that the _____ the concentration of _____, and consequently the greater the _____ , the _____ the pH value. Thus pH 4.0 indicates a _____ concentration of H^+, and a higher_____, than does pH 5.0.

Most cells are extremely sensitive to changes in the pH of their fluid. The pH of human blood plasma is usually maintained at a value between 7.34 and 7.44—that is, blood plasma is slightly _____ . The normal burning of food by the cells releases carbon dioxide (CO_2), which forms carbonic acid when combined with water. The foods we commonly eat contain sodium, potassium, and calcium, and these substances form _____ compounds within the body. When the normal limits of the blood plasma pH are greatly exceeded in either direction along the scale, _____ (pH below 6.8) or _____ (pH above 7.8) can lead to serious illness and even death, unless a proper _____ balance is restored.

Exercise 1.2

Labeling.   Write the name of each numbered body plane and the direction in the space provided. Select different colors for the separate planes—do not color the body.

Key:
dorsal, posterior
frontal plane
inferior
sagittal plane
superior
transverse plane
ventral, anterior

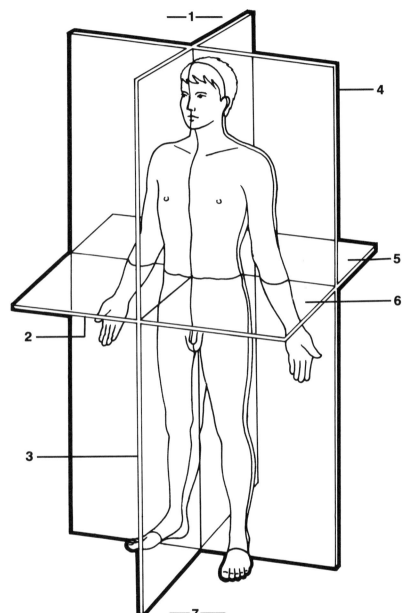

Figure 1.2   Body planes and directions.

1. _____   5. _____

2. _____   6. _____

3. _____   7. _____

4. _____

Exercise 1.3

Labeling. Write the name of each numbered body cavity or cavities in the space provided. Color the dorsal and ventral cavities differently.

Key:
abdominal cavity
cranial cavity
dorsal body cavities
pelvic cavity
spinal cord
thoracic cavity
ventral body cavity
diaphragm
vertebral column

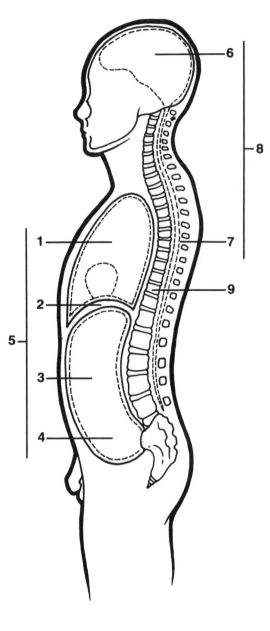

Figure 1.3 Sagittal section of the body, showing the dorsal and ventral body cavities.

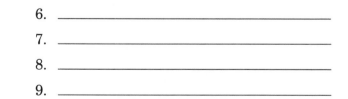

1. _____   6. _____

2. _____   7. _____

3. _____   8. _____

4. _____   9. _____

5. _____

Exercise 1.4

Labeling.   Write the name of each numbered region of the abdomen in the space provided. Select a different color for each of the nine regions.

Key:
epigastric
hypogastric
left hypochondriac
left inguinal
left lumbar
right hypochondriac
right inguinal
right lumbar
umbilical

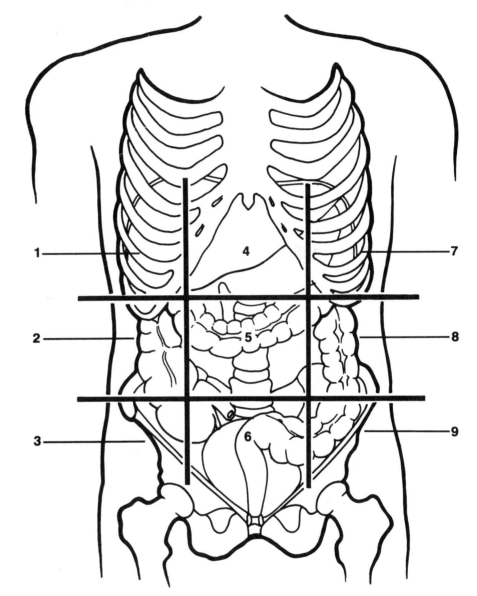

Figure 1.4   Abdominal regions.

1. _____

2. _____

3. _____

4. _____

5. _____

6. _____

7. _____

8. _____

9. _____

Exercise 1.5

The Metric System

Unit	Symbol
Centimeter	cm = 0.4 inch
Millimeter	mm = 0.1 cm
Micron (micrometer)*	μ = 0.001 mm (μm)
Millimicron (nanometer)*	mμ = 0.001 μ (nm)
Angstrom	Å = 0.1 mμ

Table 1.1 *Units of Measurement*

*The terms in parentheses have been adopted by the new international system of units of measurement to replace the ones given. However, the more familiar terms have not gone out of use as yet.

To demonstrate that you understand the relationship of one metric unit to another, fill in the blanks below.

1 mm = _____ μ 1 μ = _____ mm

1.5 mm = _____ μ 1,500 μ = _____ mm

0.25 mm = _____ μ 250 μ = _____ mm

1.5 cm = _____ mm = _____ μ

5,000 μ = _____ mm = _____ cm

V. TEST ITEMS

A. *Multiple Choice.* There is only one answer that is either correct or most appropriate. Circle the answer that corresponds to the question.

1. Which of these pH values is closer to neutrality?
   a. 6.89   c. 6.94
   b. 7.11   d. 7.05

2. On which side of the hand is the thumb?
   a. lateral   c. proximal
   b. medial   d. distal

3. The small of the back is known as the _____ region.
   a. scapular   c. epigastric
   b. sternal   d. lumbar

4. A sagittal section will cut the body into two
   a. right and left halves.
   b. superior and inferior halves.
   c. dorsal and ventral halves.
   d. anterior and posterior halves.

5. Which term describes the location of the foot in reference to the leg?
a. anterior
c. distal
b. proximal
d. posterior

6. Which term best describes the relative constancy of the interval environment of the body?
a. stress
c. homeostasis
b. pathology
d. metastasis

7. The pH of blood is slightly basic. Which of the following would be appropriate?
a. 6.4
d. 7.4
b. 4.6
e. 13.8
c. 4.7

8. The plane or section that divides the body into right and left portions is called
a. sagittal.
c. coronal.
b. transverse.
d. cross-section.

9. If you were to assume the anatomical position you would
a. lie face down.
b. lie face up.
c. stand erect with hands facing forward.
d. stand erect with thumbs backward.

10. Which term describes the location of the hand in reference to the arm?
a. anterior
c. distal
b. proximal
d. posterior

11. A cut through the body or body structure that divides it into front and back portions is
a. coronal.
c. sagittal.
b. frontal.
d. transverse.

12. Of the nine surface areas of the abdominal region, which of the following groups are medial?
a. hypochondriac, lumbar, and iliac or inguinal
b. epigastric, umbilical, and hypogastric
c. epigastric, umbilical, and iliac
d. hypochondriac, umbilical, and hypogastric

13. If you wanted to separate the abdominal from the thoracic cavity, which plane would you use?
a. sagittal
c. frontal
b. transverse
d. coronal

14. Another name for the chest cavity is the
a. ventral cavity.
c. pleural cavity.
b. thoracic cavity.
d. peritoneal cavity.

15. The peritoneum lines which cavity?
    a. abdominopelvic
    b. thoracic
    c. cardiac
    d. cranial

16. The diaphragm separates which cavities from each other?
    a. abdominal-pelvic
    b. cranial-vertebral
    c. thoracic-abdominal
    d. dorsal-ventral

17. A pulled muscle in the femoral region might affect your ability to
    a. turn your head.
    b. bend your arm.
    c. walk.
    d. move your fingers.

18. The part of the nervous system that is found within the dorsal cavity is called the
    a. central   c. autonomic
    b. peripheral   d. visceral nervous system

19. Superior is to head as lateral is to
    a. middle.   c. feet.
    b. side.   d. front.

20. Which term best describes the relationship of the elbow to the wrist?
    a. medial   c. proximal
    b. lateral   d. external

B. *Matching Questions.* Each of the phrases in COLUMN B refers to a word or phrase in COLUMN A. Insert the letter of the word or phrase from COLUMN B that best describes it. Some words or phrases may be used more than once or not at all.

Column A		*Column B*	
1. ___	proximal	**a.**	toward the feet
2. ___	inferior	**b.**	further from the point of origin
3. ___	parietal	**c.**	toward the back
4. ___	distal	**d.**	toward one side of the body
5. ___	visceral	**e.**	in close proximity to body walls
6. ___	ventral	**f.**	nearer to the point of origin
7. ___	dorsal	**g.**	toward the front
8. ___	medial	**h.**	toward the head
9. ___	superior	**i.**	next to the internal organs
10. ___	lateral	**j.**	toward the midline of the body

C. *True-False.*   Place a *T* or *F* in the space provided.

____ **1.** Cytology is the branch of microscopic anatomy that deals with the structure of tissues.

____ **2.** In the standard anatomical position, the body is erect with the feet together, the arms hanging at the sides, and the thumbs pointing away from the body.

____ **3.** The hypochondriac region of the abdomen lies below the umbilicus.

____ **4.** Sagittal and coronal planes divide the body into upper and lower parts.

____ **5.** Cranial and vertebral portions are subdivisions of the dorsal body cavity.

____ **6.** The hands are the proximal portion of the upper extremities.

____ **7.** The palms of the hands and soles of the feet are on the ventral body surface of man.

____ **8.** The numerical value of pH increases with increasing hydrogen ion concentration.

____ **9.** Because K^+, Na^+, Cl^-, and HCO_3^- all have charges, they are classified as ions.

____ **10.** The degree of acidity or alkalinity of a given solution is referred to as its pH.

Answer Sheet—Chapter 1

Exercise 1.1

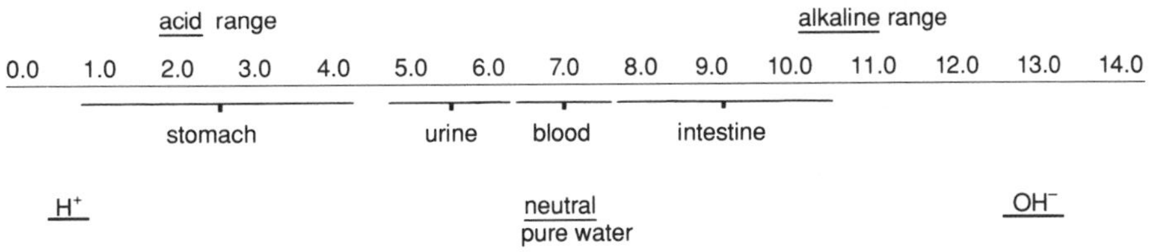

acid range							alkaline range							
0.0	1.0	2.0	3.0	4.0	5.0	6.0	7.0	8.0	9.0	10.0	11.0	12.0	13.0	14.0

stomach   urine   blood   intestine

H^+ neutral OH^-
pure water

Figure 1.1   The pH scale.

It is essential to understand that a number on the pH scale is actually the result of dividing the numeral 1 by a mathematical value called a logarithm. The result is that the higher the concentration of H^+, and consequently the greater the acidity, the lower the pH value. Thus pH 4.0 indicates a higher concentration of H^+, and a higher acidity, than does pH 5.0.

Most cells are extremely sensitive to changes in the pH of their fluid. The pH of human blood plasma is usually maintained at a value between 7.34 and 7.44—that is, blood plasma is slightly alkaline. The normal burning of food by the cells releases carbon dioxide (CO_2), which forms carbonic acid when combined with water. The foods we commonly eat contain sodium, potassium, and calcium, and these substances form alkaline compounds within the body. When the normal limits of the blood plasma pH are greatly exceeded in either direction along the scale, acidosis (pH below 6.8) or alkalosis (pH above 7.8) can lead to serious illness and even death, unless a proper acid-base balance is restored.

Exercise 1.2

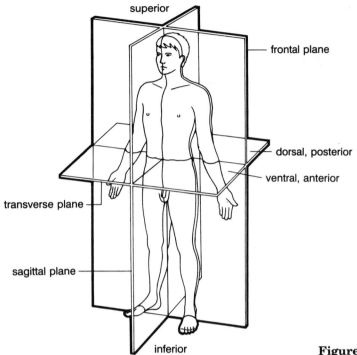

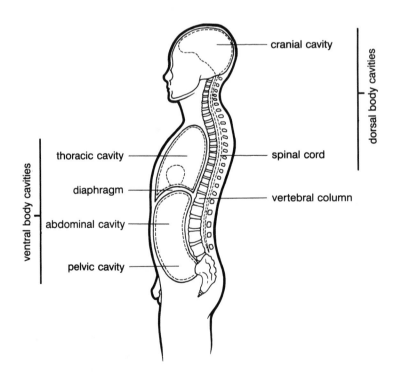

Figure 1.2   Body planes and directions.

Exercise 1.3

Figure 1.3   Sagittal section of the body, showing the dorsal and ventral body cavities.

Exercise 1.4

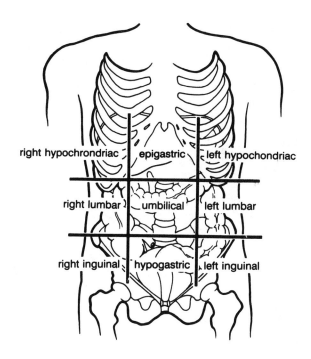

right hypochrondriac   epigastric   left hypochondriac

right lumbar   umbilical   left lumbar

right inguinal   hypogastric   left inguinal

Figure 1.4   Abdominal regions.

Exercise 1.5

1 mm   =   <u>1,000</u> μ   1 μ   =   <u>0.001</u> mm

1.5 mm   =   <u>1,500</u> μ   1,500 μ   =   <u>1.5</u> mm

0.25 mm   =   <u>250</u> μ   250 μ   =   <u>0.25</u> mm

1.5 cm   =   <u>15</u> mm   =   <u>15,000</u> μ

5,000 μ   =   <u>5</u> mm   =   <u>0.5</u> μ

Test Items

A. 1.d, 2.a, 3.d, 4.a, 5.c, 6.c, 7.d, 8.a, 9.c, 10.c, 11.a, 12.b, 13.b, 14.b, 15.a, 16.c, 17.c, 18.a, 19.b, 20.c.

B. 1.f, 2.a, 3.e, 4.b, 5.i, 6.g, 7.c, 8.j, 9.h, 10.d.

C. 1.F, 2.T, 3.F, 4.F, 5.T, 6.F, 7.T, 8.F, 9.T, 10.T.

Terminology

Across

3 closest to the origin

5 either side of the lower abdomen

6 pertaining to the backbone

9 refers to the lower back

10 pertaining to the head

13 the groin region

16 another term for forehead

17 a hydroxyl donor; high pH

20 pertaining to the back

21 excessive amount of hydroxyl ions

22 between the chest and the pelvis

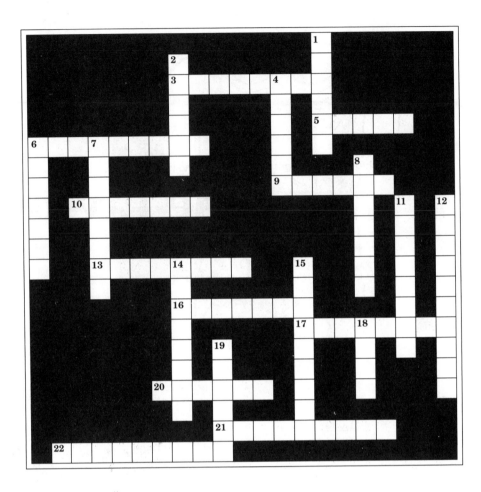

Down

1 below the abdomen

2 pertaining to the nerve cord

4 toward the middle

6 pertaining to the front

7 term for chest

8 to the side

11 excessive amount of hydrogen ions

12 across

14 below something

15 found on both sides

18 a hydrogen donor; low pH

19 further from the origin

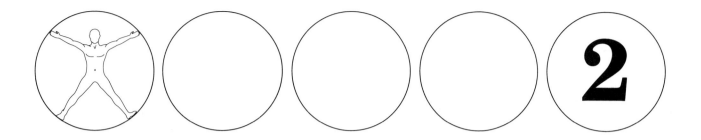

The Living Cell

I. CHAPTER SYNOPSIS

Living things or organisms display a remarkable and fundamental similarity in both structure and function. All living forms are essentially made up of one or more basic units or structural compartments called cells.

The living substance of the cell is called protoplasm. It refers to the living matter within the cell or plasma membrane. Within the protoplasm are substances or structures called organelles, each playing a major role in the total physiology of the cell.

II. OBJECTIVES

After reading the chapter, the student should be able to:

- Describe basic cellular organization.

- Give one or more important functions of the organelles.

- Describe mitochondria, ribosome, and lysosome and show their relation to cellular metabolism.

- Explain the possible roles of microtubules and microfilaments in the shape of a cell.

- Explain the complementary pairing of nucleic acids in the structure of DNA.

- Cite the main function of DNA contained in the nucleus of a cell.

- Distinguish between DNA and RNA.
- Describe the chemical and mechanical phases of mitosis.

III. IMPORTANT TERMS

Using your textbook, define the following terms:

anaphase (an'-ah-faze) _____

_____

ATP (ay-tee-pee) _____

_____

autolysis (aw-tol'-i-sis) _____

_____

autosome (awt'-o-som) _____

_____

centriole (sen'-tree-ol) _____

_____

chromatin (kro'-mah-tin) _____

_____

chromosome (kro'-mo-som) _____

_____

cisternae (sis-ter'-nee) _____

_____

codon (ko'-don) _____

_____

cytoplasm (sight'-o-plaz-em) _____

_____

DNA (dee-en-ay) _____

_____

endoplasmic reticulum (en-doe-plaz'-mik ri-tik'-yoo-lum) _____

_____

enzyme (en'-zime) _____

_____

Golgi apparatus (gol'-jee ap-ah-rat'-us) _____

_____

interphase (int'-er-faze) _____

_____

lysosome (li'-so-som) _____

_____

meiosis (mi-o'-sis) _____

_____

metaphase (met'-ah-faze) _____

_____

microfilament (mi-kro-fil'-ah-ment) _____

_____

microtubule (mi-kro-too'-byoo-el) _____

_____

mitochondria (mi-to-kon'-dree-a) _____

_____

mitosis (mi-to'-sis) _____

_____

nucleolus (nu-klee'-o-lus) _____

_____

nucleoplasm (nu'-klee-o-plaz-em) _____

_____

organelle (or-gah-nel') _____

_____

prophase (pro'-faze) _____

_____

protoplasm (prot'-ah-plaz-em) _____

_____

RNA (ar-en-ay) _____

_____

ribosome (ri'-bo-som) _____

_____

telophase (tel'-o-faze) _____

_____

transcription (trans-krip'-shun) _____

_____

translation (tranz-lay' shun) _____

_____

IV. EXERCISES

Complete the following exercises in the order given. A precise set of terms and structures has been chosen to describe the cell and DNA/RNA.

Exercise 2.1

Labeling.   Write the name of each numbered part of the cell in the space provided. Use a different color to depict a common function of each part of the cell.

Key:

centrioles

globular heads

Golgi apparatus

lipid droplets

lysosomes

mitochondrion

nuclear envelope

nucleolus

rough endoplasmic
   reticulum

secretory granules

smooth endoplasmic
   reticulum

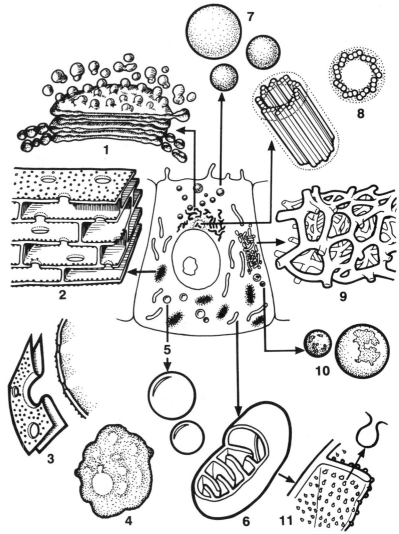

Figure 2.1   Diagram of a cell (center).

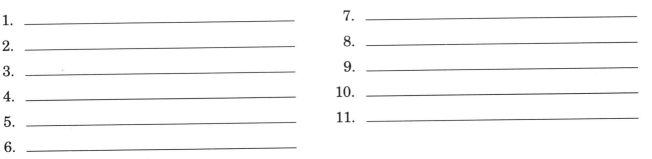

1. _____

2. _____

3. _____

4. _____

5. _____

6. _____

7. _____

8. _____

9. _____

10. _____

11. _____

Exercise 2.2

Labeling. Write the name of each numbered part of the nucleus in the space provided.

Key:
chromosomes
nucleic acids
nucleolus
nucleus

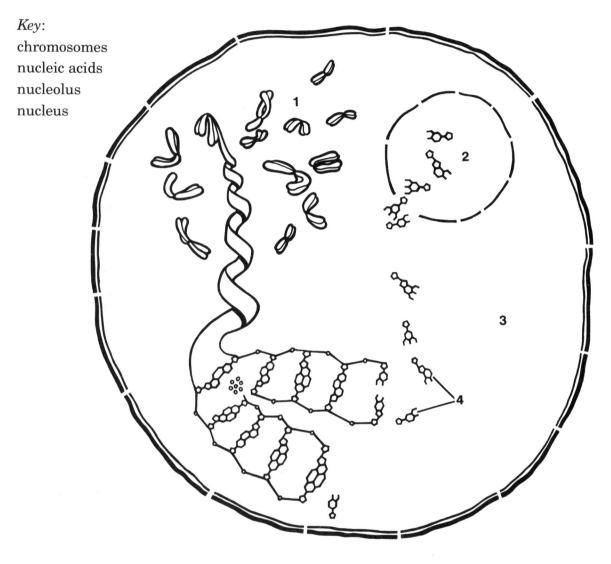

Figure 2.2 Diagram of DNA molecule, showing replication in the nucleus.

1. _____   3. _____

2. _____   4. _____

Exercise 2.3

Labeling. Write the name (symbol) of each numbered part in the space provided. Names (symbols) may be used more than once. Color the reciprocal nucleic acid pairings to match their codon amino acid.

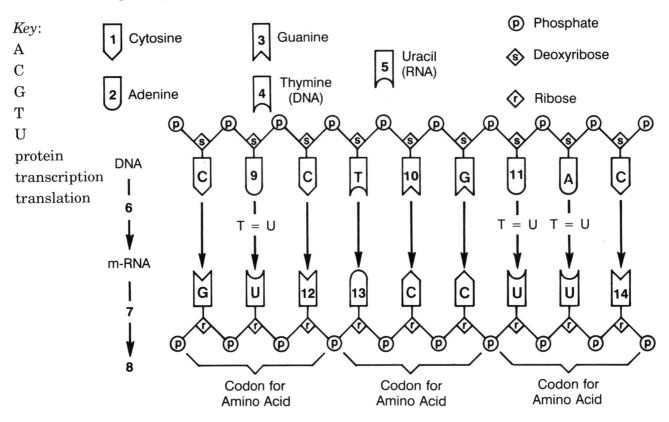

Figure 2.3   Transcription from one DNA strand to m-RNA showing codons that specify amino acids. Phosphates connect ribose molecules. Abbreviations: A, adenine; T, thymine; G, guanine; C, cytosine; R, ribose sugar; U, uracil; P, phosphate radicals.

1. _____
2. _____
3. _____
4. _____
5. _____
6. _____
7. _____
8. _____
9. _____
10. _____
11. _____
12. _____
13. _____
14. _____

Exercise 2.4

Labeling.   Write the name of the part or phase of mitosis of a living cell.
Color the nuclear components differently from the cytoplasm.

Key:
anaphase
cell membrane
centrioles
chromosomes
early prophase
early telophase
interphase
interphase
late prophase
late telophase
metaphase
nuclear membrane

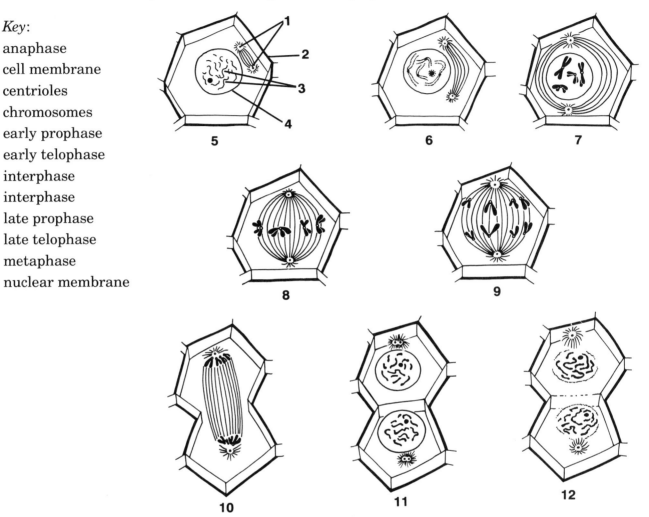

Figure 2.4   Schematic representation of mitosis of a cell containing four
chromosomes. See text for an account of what happens during each of the various
stages.

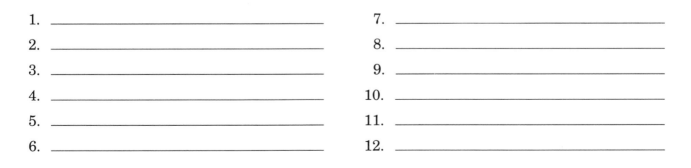

1. _____

2. _____

3. _____

4. _____

5. _____

6. _____

7. _____

8. _____

9. _____

10. _____

11. _____

12. _____

V. TEST ITEMS

A. *Multiple Choice.*   There is only one answer that is either correct or most appropriate. Circle the answer that corresponds to the question.

1. Autolysis, or self-destruction, of cells would be caused by enzymes in
   a. the smooth endoplasmic reticulum.
   b. lysosomes.
   c. mitochondria.
   d. none of the foregoing

2. Most of the cell's ATP is manufactured in organelles called
   a. mitochondria.   c. ribosomes.
   b. Golgi apparatus.   d. lysosomes.

3. Which of the following organelles "packages" substances synthesized by the cell?
   a. Golgi apparatus   c. lysosomes
   b. mitochondria   d. ribosomes

4. Which of the following is the genetic matter of the cell?
   a. DNA   c. proteins
   b. RNA   d. none of the foregoing

5. Which of the following is used by the cell as the direct source of energy to carry out its various processes?
   a. glucose   c. ATP
   b. fatty acids   d. protein

6. The site of protein synthesis is
   a. ribosomes.
   b. smooth endoplasmic reticulum.
   c. lysosomes.
   d. none of the foregoing

7. Which of the following become the "poles" of the cell to which chromosomes migrate during cell division?
   a. ribosomes   c. lysosomes
   b. mitochondria   d. centrioles

8. The endoplasmic reticulum
   a. functions as an extracellular network.
   b. attaches to chromosomes during division.
   c. forms the cleavage furrow during division.
   d. serves as an internal framework and an intracellular passageway.

9. As a result of cell division, each daughter cell has
   a. half as many chromosomes as its parent cell.
   b. twice as many chromosomes as its parent cell.
   c. just as many chromosomes as its parent cell.
   d. one-quarter as many chromosomes as its parent cell.

10. A cell whose chromosomes are migrating to opposite poles is a cell in
   a. prophase.   c. metaphase.
   b. anaphase.   d. telophase.

11. The organelle associated with synthesis, storage, and secretion of glycoproteins is in the
    a. nucleus.
    b. Golgi complex.
    c. mitochondrion.
    d. lysosome.

12. The linkage of bases in DNA follows a pattern in which a purine base is always linked with a pyrimidine base. Possible linkages are
    a. A-G.
    b. A-T.
    c. C-U.
    d. T-C.
    e. C-T.

13. The genetic material in a nondividing cell is referred to as
    a. chromatin.
    b. centrioles.
    c. chromosomes.
    d. karyolymph.

14. The nucleus of a cell
    a. contains Golgi bodies.
    b. lacks a nuclear membrane.
    c. contains chromatin.
    d. contains a centrosome.
    e. lacks nucleoplasm.

15. The unit of structure and function of living things is
    a. protoplasm.
    b. the cell.
    c. an organ.
    d. a nucleus.

16. Microtubules
    a. are composed of protein.
    b. are found in centrioles, cilia, and flagella.
    c. may be associated with movement.
    d. are found in the cytoplasm.
    e. all of these

17. Select the base that is not always found in both DNA and RNA.
    a. adenine
    b. uracil
    c. guanine
    d. cytosine

18. One kind of base pair in DNA molecules is adenine joined to
    a. cytosine.
    b. guanine.
    c. thymine.
    d. deoxyribose.
    e. ribose.

19. Rough endoplasmic reticulum is mostly associated with
    a. ribosome and protein formation.
    b. mitochondria and respiration.
    c. centrosome and mitosis.
    d. ribosome and respiration.

20. Separation of the daughter cells occurs in
    a. prophase.
    b. interphase.
    c. metaphase.
    d. telophase.
    e. anaphase.

B. *Matching Questions.*   Each of the phrases in COLUMN B refers to a word or phrase in COLUMN A. Insert the letter of the word or phrase from COLUMN B that best describes it. Some words or phrases may be used more than once or not at all.

Column A

1. ___ synthesis of ATP
2. ___ package and storage of cellular secretions
3. ___ removal of unwanted cellular substances
4. ___ synthesis of protein
5. ___ synthesis of mitotic spindle
6. ___ storage of solid or fluid substance
7. ___ conversion of proteins to glycoprotein
8. ___ channels for passage of substance within the cell
9. ___ transmission of genetic information
10. ___ synthesis of ribosomal RNA

Column B

a. Golgi apparatus
b. nucleus
c. mitochondria
d. ribosomes
e. microtubules
f. centrioles
g. lysosomes
h. chromosomes
i. vacuoles
j. endoplasmic reticulum (ER)

Column A

1. ___ chromosomes
2. ___ centrioles reach poles; line up on equator
3. ___ genetic information replicated
4. ___ nuclear membrane disappears
5. ___ two daughter cells are produced
6. ___ chromatin becomes organized

Column B

a. prophase
b. interphase
c. telophase
d. anaphase
e. metaphase

C. *True-False.*   Place a *T* or *F* in the space provided.

___ 1. The science that deals with the study of cells is cytology.

___ 2. In DNA and RNA the backbone of the nucleic acid chain consists of alternating units of phosphate and a 5-carbon sugar.

___ 3. The nuclear envelope disintegrates during metaphase.

___ 4. The nuclear membrane forms around each group of daughter chromosomes during telophase.

___ 5. In DNA the two chains are joined by bonding between specific base pairs—adenine to thymine, guanine to cytosine.

___ **6.** Parts of the cell that are specialized for specific activities are called organelles.

___ **7.** The portion of the cell that controls its activities and contains hereditary material is the nucleus.

___ **8.** The 5-carbon sugar in DNA is called ribose.

___ **9.** The mitochondria are called "powerhouses" of the cell because ATP is produced within them.

___ **10.** Water constitutes the most abundant single compound found in the human body.

Answer Sheet—Chapter 2

Exercise 2.1

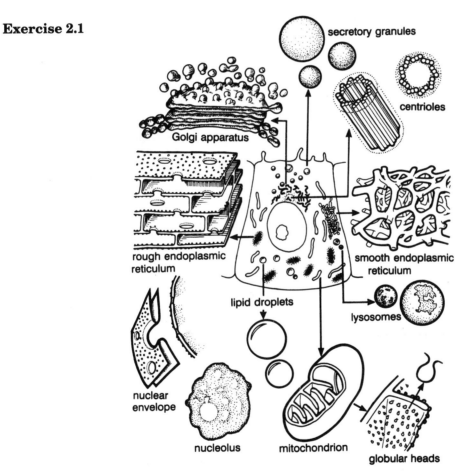

Figure 2.1 Diagram of a cell (center).

Exercise 2.2

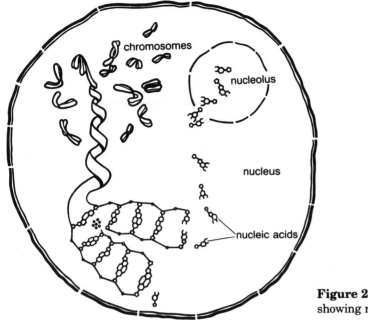

Figure 2.2   Diagram of DNA molecule, showing replication in the nucleus.

Exercise 2.3

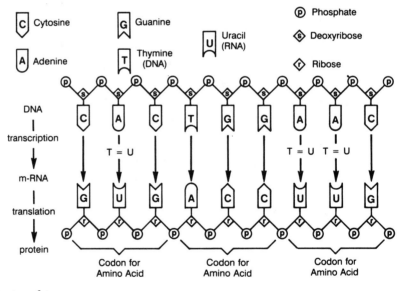

Figure 2.3   Transcription from one DNA strand to m-RNA showing codons that specify amino acids. Phosphates connect ribose molecules. Abbreviations: A, adenine; T, thymine; G, guanine; C, cytosine; R, ribose sugar; U, uracil; P, phosphate radicals.

Exercise 2.4

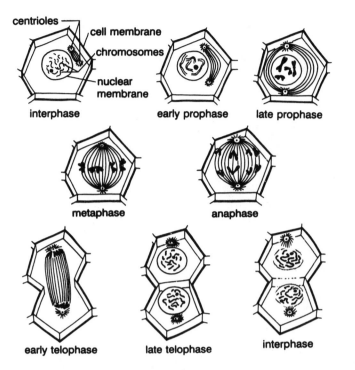

Figure 2.4   Schematic representation of mitosis of a cell containing four chromosomes. See text for an account of what happens during each of the various stages.

Test Items

A.   1.b, 2.a, 3.a, 4.a, 5.c, 6.a, 7.d, 8.d, 9.c, 10.b, 11.b, 12.b, 13.a, 14.c, 15.b, 16.e, 17.b, 18.c, 19.a, 20.d.

B.   1.c, 2.a, 3.g, 4.d, 5.f, 6.i, 7.a, 8.j, 9.h, 10.b.
     1.d, 2.e, 3.b, 4.a, 5.c, 6.b.

C.   1.T, 2.F, 3.F, 4.T, 5.T, 6.T, 7.T, 8.F, 9.T, 10.T.

The Cell

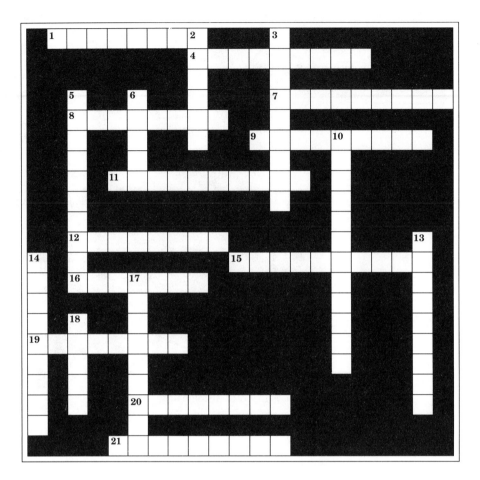

Across

1 chromosome without sex traits

4 nuclear organelle that stores nucleic acids

7 self-digestion of a cell

8 organelle that produces proteins

9 nuclear material that forms chromosomes

11 biochemical phase of cell division

12 chromosomes move to opposite poles

15 contains the hereditary material

16 sexual cell division

19 chromosomes are visible and appear duplicated

20 cellular organelle that stores enzymes

21 organelle that produces spindle fibers

Down

2 an organic catalyst in metabolism

3 chromosomes line up across the middle of the cell

5 the living substance of the cell

6 three nucleotides of an RNA molecule

10 the powerhouse of the cell

13 the end phase of cell division

14 protoplasm of a cell exclusive of the nucleus

17 microscopic functioning structure of a cell

18 a secretory cellular organelle

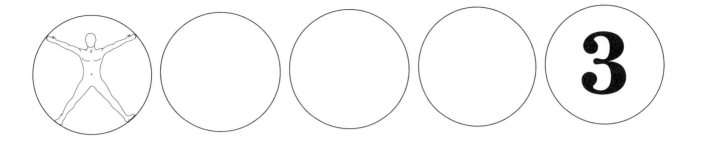

The Cell Membrane and Permeability

I. CHAPTER SYNOPSIS

The plasma cell membrane is a delicately balanced, functional organelle that separates the cell from its environment and allows materials to pass across it in both directions. By controlling ionic composition and water content, the membrane prevents the cell from swelling or shrinking and, therefore, is directly responsible for maintaining homeostasis.

The main barrier to substance exchange across the membrane is undoubtedly the lipid layer. A molecule must be able to pass through the small pores of the cell membrane or dissolve in the lipid layer and then diffuse through the membrane. Water passes readily through all membranes. Small, positively charged ions, however, move through the membrane very slowly due to the postulated positive charge of the lipid pores. Like charges repel each other. Negatively charged ions pass more easily.

II. OBJECTIVES

After reading the chapter, the student should be able to:

- Describe the molecular structure of a cell membrane.
- Define semipermeable membrane in terms of its role in governing the exchange of chemical compounds between cell and compartment.
- Define diffusion and osmosis in terms of movement of particles or liquids in response to a concentration gradient.

- Differentiate between facilitated diffusion and active transport.
- Define phagocytosis and pinocytosis.
- Define isotonic, hypotonic, and hypertonic and their effects on cell volume.

III. IMPORTANT TERMS

Using your textbook, define the following terms:

active transport (ack'-tiv tranz-port') _____

_____

autolysis (aw-tol'-i-sis) _____

_____

concentration gradient (kon-sun-tray'-shun gray'-dee-unt) _____

_____

crenation (kre-nay'-shun) _____

_____

dehydration (de-hi-dray'-shun) _____

_____

dialysis (di-al'-i-sis) _____

_____

diffusion (dif-yoo'-zhun) _____

_____

equilibrium (ee-kwi-lib'-ree-um) _____

_____

filtrate (fil'-trate) _____

_____

filtration (fil-tray'-shun) _____

_____

glycocalyx (gli-ko-kay'-liks) _____

_____

hemolysis (hee-mah-li'-sis) _____

_____

homeostasis (ho-mee-o-stay'-sis) _____

_____

hydrophilic (hi-dro-fil'-ik) _____

_____

hydrophobic (hi-dro-fo'-bik) _____

_____

hypertonic (hi-per-tahn'-ik) _____

_____

hypotonic (hi-po-tahn'-ik) _____

_____

isotonic (i-so-tahn'-ik) _____

_____

lipid (lip'-id) _____

_____

mosaic (mo-zay'-ik) _____

_____

osmosis (oz-mo'-sis) _____

_____

permeability (per-mee-ah-bil'-i-tee) _____

_____

permeable (per'-mee-ah-bul) _____

_____

phagocytosis (fag-o-si-tow'-sis) _____

_____

pinocytosis (pin-o-si-tow'-sis) _____

_____

pore (poor) _____

_____

protein (pro'-teen) _____

_____

semipermeable (sem-ee-per'-mee-ah-bul) _____

_____

solute (sol'-yoot) _____

_____

solvent (sol'-vent) _____

_____

IV. EXERCISES

Complete the following exercises in the order given. A precise set of terms and structures has been chosen to describe the cell membrane and permeability.

Exercise 3.1

Labeling. Write the name of each numbered part of the diagram in the space provided. Color the different parts to identify their position in the membrane.

Key:

carbohydrate receptor sites (glycocalyx)

enzyme

hydrophilic layer

hydrophobic layer

identity

phospholipid

pore

protein molecules

transport

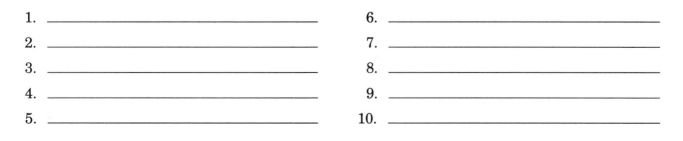

Figure 3.1   Fluid-mosaic model of the cell membrane.

1. _____
2. _____
3. _____
4. _____
5. _____

6. _____
7. _____
8. _____
9. _____
10. _____

Exercise 3.2

Labeling.   Write the name of each numbered part of the diagram in the space provided. Color the direction of flow in each instance. Terms may be used more than once.

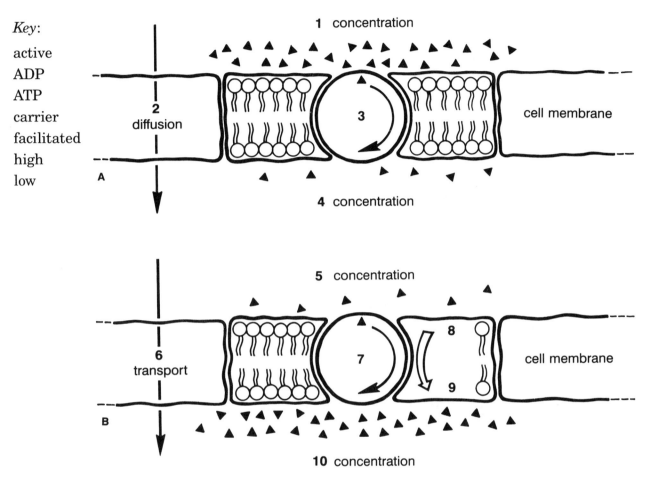

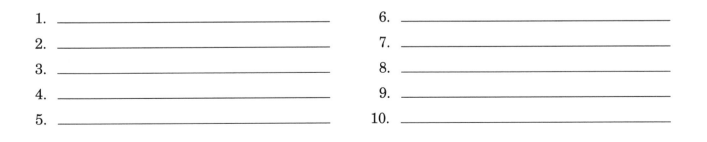

Figure 3.2   Diagram illustrating two-carrier molecular system. **A.** Movement of molecule (△) across the membrane from higher to lower concentration of molecule. **B.** Movement of a molecule (△) across a membrane from low to high concentration of molecule with the expenditure of energy to afford the transport.

1. _____   6. _____

2. _____   7. _____

3. _____   8. _____

4. _____   9. _____

5. _____   10. _____

V. TEST ITEMS

A. *Multiple choice.* There is only one answer that is either correct or most appropriate. Circle the answer that corresponds to the question.

1. Carrier molecules are required for
   a. diffusion.
   b. osmosis.
   c. facilitated transport.
   d. active transport.
   e. both c and d

2. The cell membrane is composed of
   a. cellulose.
   b. cellulose and protein.
   c. entirely of lipid.
   d. lipid and protein.

3. Permeability is
   a. the movement of molecules from the area of greater concentration to the area of lesser concentration.
   b. the ability to allow a substance to pass through.
   c. the state of being permanent.
   d. a term that is only applicable to membranes.

4. Which of these is absolutely necessary for diffusion to take place?
   a. a differentially permeable membrane
   b. a true solution
   c. a living cell
   d. a concentration difference
   e. a permeable membrane

5. When a substance moves from an area of low concentration to an area of high concentration
   a. diffusion has occurred.
   b. the cell bursts.
   c. energy is needed.
   d. osmotic pressure builds up.

6. Proteins do not pass through cell membranes because
   a. the membrane is made of protein.
   b. they contain nitrogen.
   c. they are very large molecules.
   d. they cause emulsification.

7. If 0.9 percent NaCl were isotonic to a cell, then
   a. 0.9 percent would also be hypotonic.
   b. 0.9 percent would also be hypertonic.
   c. 1.0 percent would be hypertonic.
   d. 1.0 percent would be hypotonic.

8. Pinocytosis and phagocytosis are accomplished by the
   a. nucleus.
   b. mitochondria.
   c. cell membrane.
   d. endoplasmic reticulum.

9. Isotonic means
   a. the effective concentration of the dissolved substances in surrounding fluid is greater than the concentration in the cell.
   b. the effective concentration of the surrounding fluid is less than the concentration in the cell.
   c. the effective concentration of the dissolved substances in the surrounding fluid is the same as the concentration within the cell.
   d. all of these are true

10. A cell in a hypotonic solution
    a. loses water.
    b. gains water.
    c. neither gains nor loses water.
    d. both gains and loses water equally.

11. Osmosis occurs when a membrane is
    a. impermeable.
    b. differentially permeable.
    c. permeable.
    d. both a and c

12. The fact that lipids move easily through cell membranes is due to
    a. their size.
    b. their chemical composition.
    c. osmosis.
    d. active transport.

13. Cells will swell in
    a. a hypertonic solution.
    b. a hypotonic solution.
    c. an isotonic solution.
    d. none of the foregoing

14. In which process is liquid "pushed" through a semipermeable membrane or filter from an area of higher pressure into an area of lower pressure?
    a. osmosis
    b. active transport
    c. filtration
    d. diffusion

15. During diffusion, a substance always moves from a region
    a. outside a cell to the inside of a cell.
    b. inside of a cell to the outside of a cell.
    c. of higher concentration to a region of lower concentration.
    d. of lower concentration to a region of higher concentration.

16. The separation of small molecules from large ones by diffusion of the smaller molecules through a semipermeable membrane is known as
    a. osmosis.
    b. filtration.
    c. active transport.
    d. dialysis.

17. In the body, the proper concentration and distribution of various inorganic salts is referred to as
    a. homeostasis.
    b. inorganic equilibrium.
    c. ionic stability.
    d. electrolyte balance.

18. If red blood cells are placed in a hypertonic solution of sodium chloride, what will happen to them?
    a. They will shrink.
    b. They will swell.
    c. They will burst.
    d. They will stick together.

19. The plasma membrane
    a. is a semipermeable membrane.
    b. surrounds the nucleus.
    c. contains cellulose.
    d. forms and excretes calcium.

20. The assimilation of fluid into a cell occurs because of
    a. phagocytosis.
    b. pinocytosis.
    c. passive transport.
    d. active transport.

B. *Matching Questions.*   Each of the phrases in COLUMN B refers to a word or phrase in COLUMN A. Insert the letter of the word or phrase from COLUMN B that best describes it. Some words or phrases may be used more than once or not at all.

Column A		*Column B*
1. ___ solute		**a.** parts/volume
2. ___ solvent		**b.** greater to lesser
3. ___ concentration gradient		**c.** kinetic energy
4. ___ homeostasis		**d.** particle or substance
5. ___ random movement		**e.** solution or vehicle
		f. dynamic equilibrium

Column A		*Column B*
1. ___ osmosis		**a.** enzyme-assisted diffusion
2. ___ diffusion		**b.** energy-assisted transport
3. ___ filtration		**c.** particle from greater to lesser
4. ___ facilitated transport		**d.** fluid from greater to lesser
5. ___ active transport		**e.** forced diffusion

C. *True-False.*   Place a *T* or *F* in the space provided.

____ **1.** During diffusion there is a net movement of substances from low to high concentration of the diffusing substance.

____ **2.** During osmosis there is a net movement of water from the dilute to the more concentrated solution.

____ **3.** Osmosis refers to the scattering or spreading out of molecules.

____ **4.** In active transport, liquid is literally "pushed" through a semipermeable membrane from an area of higher pressure to an area of lower pressure.

____ **5.** Active transport involves movement of ions against the concentration gradient.

____ **6.** Molecular size and solubility influence the transfer of molecules.

____ **7.** All the molecules of the same size diffuse through the membrane at the same rate.

____ **8.** The final equilibrium state reached by a molecule undergoing facilitated diffusion is the same as that for a molecule undergoing simple diffusion.

____ **9.** The principle of dialysis is employed in the operation of an artificial kidney.

____ **10.** In osmosis the molecules of solvent pass through a membrane in only one direction.

Answer Sheet—Chapter 3

Exercise 3.1

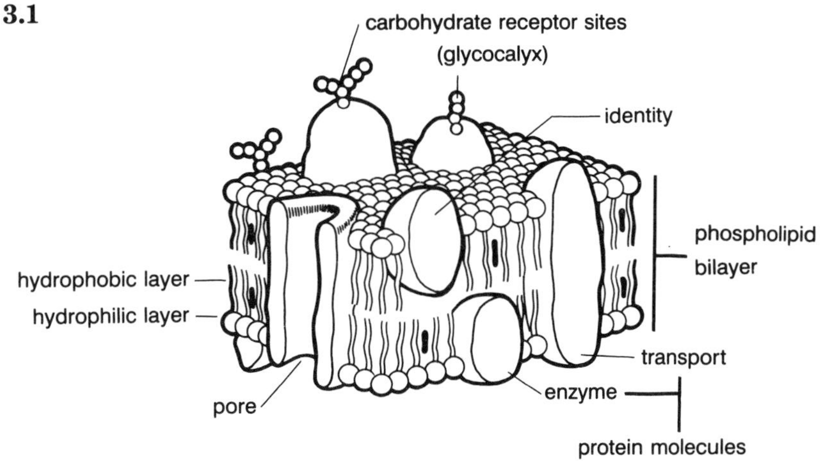

Figure 3.1 Fluid-mosaic model of the cell membrane.

Exercise 3.2

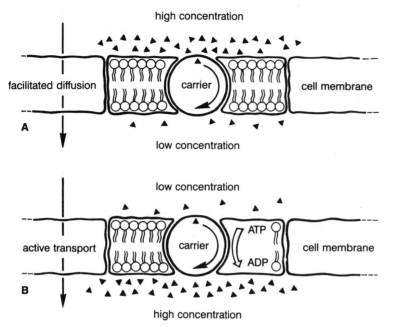

Figure 3.2 Diagram illustrating two-carrier molecular system. **A**. Movement of molecule (Δ) across the membrane from higher to lower concentration of molecule. **B**. Movement of a molecule (Δ) across a membrane from low to high concentration of molecule with the expenditure of energy to afford the transport.

Test Items

A. 1.e, 2.d, 3.b, 4.d, 5.c, 6.c, 7.c, 8.c, 9.c, 10.b, 11.b, 12.b, 13.b, 14.c, 15.c, 16.d, 17.a, 18.a, 19.a, 20.b.

B. 1.d, 2.e, 3.a, 4.f, 5.c.
    1.d, 2.c, 3.e, 4.a, 5.b.

C. 1.F, 2.T, 3.F, 4.F, 5.T, 6.T, 7.F, 8.T, 9.T, 10.F.

Permeability

Across

1 greater salt concentration outside the cell

4 shrinkage of a cell

6 movement of materials through a membrane due to liquid force

7 excessive loss of water

9 liquid that dissolves a substance

10 pattern of small particles

12 carbohydrate outer covering of a cell

14 a fatty substance

15 liquid that has passed through a membrane

16 consistency and uniformity of an environment

17 cell engulfing solid material

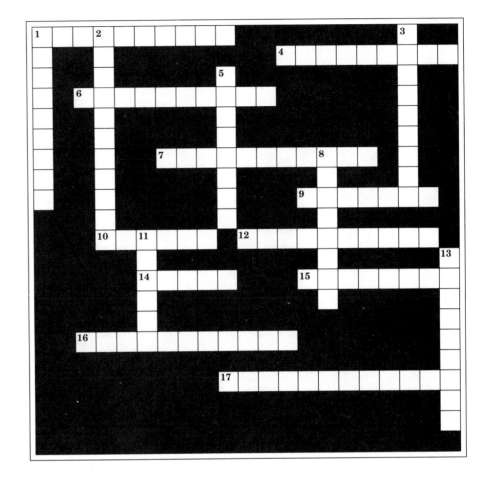

Down

1 lesser salt concentration outside the cell

2 a dynamic balance

3 movement of particles from greater to lesser concentration

5 separation of larger particles from smaller particles

8 equal salt concentration

11 substance to be dissolved

13 rupture of blood cell

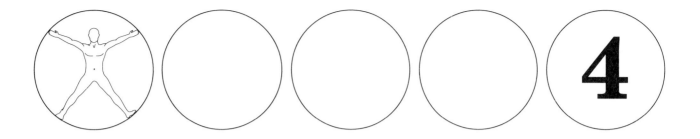

The Tissues and Integument

I. CHAPTER SYNOPSIS

The primary concern of this chapter is the organization of cells into tissues. The structure, function, and location of the principal kinds of epithelium and connective tissues are examined. Throughout, the relationship of structure to function is emphasized. The student is introduced to the organ and the system levels of organization by considering the structure and functions of the skin and its derivatives.

The integument, or skin, is a vital organ, serving as a protective barrier that responds to both internal and external challenges and contributes to the maintenance of homeostasis. Structurally, the skin is a complex combination of tissues consisting of two layers, the epidermis and dermis. The epidermis, a thin surface layer, is firmly cemented to the dermis, the deeper layer of skin that lies below it. The dermis, in turn, is attached through the subcutaneous tissue or superficial fascia to underlying structures such as bones and muscles. The appendages of the skin (the hair, cutaneous glands, and nails) develop embryologically from epidermal cells that migrate into the dermis. The blood supply to the skin nourishes the tissue cells and helps regulate body temperature. The nerve supply of the skin involves afferents from cutaneous receptors to the central nervous system (CNS) and efferents to smooth muscle. The skin performs the essential functions of protection, temperature regulation, sensation of external stimuli, production of vitamin D, and, to a minor extent, elimination of water and salts.

II. OBJECTIVES

After reading the chapter, the student should be able to:

- Differentiate between the kind of epithelial and connective tissues.
- Explain the histological anatomy of the skin and its accessory structures.
- Identify the three types of skin burns with their respective skin strata.
- Detail the steps in wound healing.

III. IMPORTANT TERMS

Using your textbook, define the following terms:

benign (be-nine') _____

_____

cancer (kan'-sur) _____

_____

carcinoma (kahr-si-no'-mah) _____

_____

collagen (kol'-ah-jen) _____

_____

connective (ko-nek'-tiv) _____

_____

corneum (kor'-nee-um) _____

_____

dermis (der'-mis) _____

_____

endocrine (en'-do-krin) _____

_____

epidermis (ep-i-der'-mis) _____

_____

epithelium (ep-i-thee'-lee-um) _____

_____

exocrine (ek'-so-krin) _____

_____

gland (gland) _____

_____

integument (in-teg'-u-ment) _____

_____

keratin (ker'-ah-tin) _____

_____

lacuna (lah-ku'-nah) _____

_____

malignant (mah-lig'-nant) _____

_____

matrix (may'-triks) _____

_____

melanin (mel'-ah-nin) _____

_____

neoplasm (nee'-ah-plaz-em) _____

_____

papilla (pah-pil'-ah) _____

_____

sarcoma (sahr-ko'-mah) _____

_____

stratum (strat'-um) _____

_____

tissue (tish'-u) _____

_____

IV. EXERCISES

Complete the following exercises in the order given. A precise set of terms and structures has been chosen to describe the cell membrane and integument.

Exercise 4.1

Completion.   Fill in the blanks with the appropriate terms.

Key:

abut

basement

continuous

covers

lines

outside

sheet

stratified

EPITHELIUM

Epithelial tissue is a complex protective layer that _____ the body and _____ all the cavities and organs having a direct connection to the _____ of the body. To put it simply, epithelium is a kind of _____ made up of a series of cells that _____ each other. Epithelial cells are so closely joined together at these junctions that they form a _____ barrier between the body parts they cover and the surrounding medium (water, air, or internal body fluids).   Epithelial tissue may be composed of only one flat layer of cells, or it may be _____ into different layers. When examined under a microscope, epithelial cells can be seen to have a supporting _____ membrane, which appears in prepared sections as a fine line.

Key:

carbon dioxide

exchange

irregular

oxygen

single

scalelike

shapes

Types of Epithelial Tissue

Four distinct types of epithelium occur in the human body. Their classification is based on the _____ and properties of the cells composing them.

1. *Squamous epithelium.* Simple squamous epithelium is made up of flat cells arranged in a _____ layer. The term squamous means _____, and the cells of these tissues have _____ shapes like scales. Each cell contains a large, prominent nucleus at its center. In its simplest form, squamous epithelial tissue lines the small saclike structures of the lung (alveoli), which function in the _____ of _____ and _____ during breathing. The lens of the eye is also made up of simple squamous epithelium.

Key:

cube-shaped

glands

kidney

secretions

2. *Cuboidal epithelium.* As its name suggests, cuboidal epithelium is made up of _____ cells. This tissue lines the ducts of many _____ and the _____. The nuclei of its cells are spherical and usually are found in the center of the cell. Some cuboidal cells are capable of forming _____ and consequently are found in glands such as the thyroid, sweat glands, and salivary glands.

Key:

absorption

basement

columnlike

digestive

secretion

tall

Key:

arranged

bronchi

cilia

currents

digestion

empty

flask

goblet

hairlike

layers

mucoid

mucus

passage

respiratory

shape

surface

tall

trachea

wavelike

Key:

capillaries

circulatory

directly

endocrine

internal

no

passages

3. *Columnar epithelium.* Columnar epithelial cells are compressed to form _____ shapes. These _____ thin cells have a nucleus that usually can be found near the _____ membrane. Epithelial tissues made up of these cells occur in the _____ tract, particularly in the intestines. Columnar epithelium is concerned primarily with the _____ of digestive fluids and with the _____ of food materials.

4. *Pseudostratified epithelium.* This fourth type of epithelium earned its name because on first glance it appears to be _____ in _____ . This appearance is caused by variations in the _____ of each cell making up the tissue. Although some of the cells in contact with the basement membrane do not reach the _____ of the tissue, most of the cells are _____ and do reach the surface. Pseudostratified epithelium is found most often in the _____ tract, particularly in the _____ and in the _____ of the lung.

Two modifications occurring in the cells of epithelial tissues are worth considering here: _____ cells and _____ . Goblet cells are _____ shaped and contain _____ secretion. Microscopically, they appear as open, "_____" cells in their tissues. Cilia are _____ appendages of the cell. With their continuous _____ motion, they produce _____ in the fluids at the cell's surface. Both goblet cells and cilia occur in columnar and pseudostratified epithelium. Goblet cells aid _____ by secreting _____ for absorption of partially digested foods; cilia aid in the _____ of food down the intestinal tract. Goblet cells and cilia are also essential in the respiratory tract, where goblet cells add moisture to the air taken in and cilia clean the air of foreign particles that could otherwise clog the alveoli of the lungs.

Glands

Some epithelial tissues are made up of cells specifically organized to cause secretion.

_____ glands are sometimes called glands of _____ secretion because they are situated far beneath the epithelial surface and have _____ ducts or _____ by which their secretions can pass through the epithelium. Instead, they communicate _____ with the _____ system through the _____, which permit the distribution of their secretions throughout the body. The thyroid is an example of an endocrine gland.

Key:

duct

ducts

endocrine

enzymes

exocrine

intestine

surface

Key:

adjacent

binding

compositions

framework

matrix

physical

protection

separated

storage

Key:

fibroblasts

fibrocyte

intercellular

lose

mesenchyme

Like the endocrine glands, _____ glands are also located away from the epithelial surface. However, the exocrine glands are equipped with _____ that carry their secretions to the tissue _____ . The salivary glands, for example, are exocrine glands. The pancreas is a gland that is both endocrine and exocrine. It produces digestive _____ that are passed to the _____ through the pancreatic _____ (exocrine); but also produces a hormone, insulin, which is transported through the body by the circulatory system (_____).

CONNECTIVE TISSUES

Connective tissues are found throughout the body. As their name implies, the principal function of these tissues is _____ the body parts together. They form a _____ for the internal organs; they also perform a variety of other functions, ranging from _____ against injury to _____ of fat.

A fundamental difference between connective tissue and epithelial tissue can be seen in their cellular _____ . Epithelial cells are directly _____ to one another, separated only by a very small amount of intercellular substance called _____ . On the other hand, connective tissue contains few cells, and these are widely _____ . The intercellular matrix is relatively abundant and usually determines the _____ characteristics of a given connective tissue.

Connective Tissue Cells

In their embryonic stage, typical connective tissue cells are large and star shaped, with many projections called processes. These cells are _____ , and they arise from an early embryonic tissue, the _____ . This tissue develops into many different forms, including blood cells and muscle. As the connective tissues develop, the cells _____ their star-shaped appearance and become widely separated by large amounts of _____ material. The adult connective tissue cell, which is responsible for the formation of fibers, is the _____ . Other types of cells found in connective tissue are:

Key:
allergic
antibodies
anticoagulant
bacteria
defense
destroy
foreign
heparin
histamine
ingesting
phagocytosis
signet
store

Key:
bone
cartilage
cobweb
collagen
collagenous
elastic
elastic
fibers
glands
intercellular
ligaments
nonelastic
reticular
reticulum
tendons
yellow

1. *Histiocytes or macrophages.* These cells move about through the connective tissue, _____ _____ materials, bacteria, and cellular debris (_____).

2. *Plasma cells.* These small, irregular cells are associated with the formation of _____ , an important part of the body's _____ against foreign substances.

3. *Mast cells.* These are located near the blood vessels and are involved in the production of _____, an _____. They are also important in the production of _____ in _____ reactions.

4. *Blood cells.* The white blood cells, such as lymphocytes, monocytes, and neutrophils, are often present in connective tissue, where their function is to _____ _____ by phagocytosis.

5. *Fat cells.* These specialized cells _____ fats and oils. Microscopically, a fat cell resembles a _____ ring because the stored fat pushes the nucleus and the cytoplasm to one side of the cell.

Connective Tissue Fibers

The _____ characteristic of connective tissue are found within the _____ matrix. There are three general types:

1. _____ *fibers.* These white fibers contain _____ , an albumin that is the principal structural protein of the body. This protein is synthesized by fibrocytes and is the major constituent of skin, _____ , _____ , _____ , and _____ . The fibers occur in bundles and are relatively nonelastic.

2. _____ *fibers.* These fibers are _____ and may occur singly or in bundles. They are highly _____ and branch to unite with other fibers. Elastic fibers are both larger and straighter than collagenous fibers.

3. _____ *fibers.* These short and very thin fibers branch freely, forming a _____ network called a _____. These fibers are _____ and are usually found forming the internal framework of _____ .

Key:

adipose	loosely
areolar	padding
fat	reserve
heat	supply
insulate	

Types of Connective Tissue

Adipose tissue. _____ tissue beneath the epidermis is commonly filled with _____ packed _____ cells held together in bundles by collagenous and elastic fibers. _____ (fat) tissue is found primarily as a kind of _____ around the joints, as soft pads between the organs, around the kidney and the heart, and in the yellow marrow of long bones. Its fat cells provide a _____ food _____ and also _____ the body against _____ loss.

Key:

aponeuroses	muscles
bones	restrain
cover	separate
enclose	stronger
fasciae	tendons
ligaments	
muscle	

Dense connective tissue. Many parts of the body require a _____ type of connective tissue to _____ , _____ , or _____ functioning structures.

The basic forms of dense connective tissue are _____, which attach _____ to bone; _____ , which connect the _____ that form joints; _____ , which are thin, tendinous sheets attached to flat _____; and _____, the thin sheets of tissue that _____ muscles and hold them in place.

Key:

capsules
chondrocytes
covers
elastic
fibrous
hyaline
intervertebral
lacunae
modified
move
reinforces
rigid
skeleton

Cartilage. Cartilage is a _____ form of connective tissue in which cartilage cells (_____) lie in small _____ , the _____ , which are surrounded by an irregular matrix made up of fibers and/or a gel.

_____ cartilage, the most common form of cartilage, is modified areolar tissue with a white or glassy appearance. It may contain any number of unevenly dispersed lacunae. Hyaline cartilage forms the _____ in the embryo and _____ the articulating surfaces of bones in the joint cavities. It also provides the structure for the nose and the connections of the ribs to the breast bone, and, in the respiratory tract, forms the ringlike trachea and bronchi.

_____ cartilage contains many elastic fibers and is found wherever cartilage is required to _____ , such as in the epiglottis and the external ear.

_____ cartilage is less _____ than hyaline, but contains heavy bundles of collagenous connective tissue. It is found in the _____ disks, which absorb shocks between the vertebrae of the backbone. Fibrous cartilage also _____ the hyaline articular cartilages at the knee and hip.

Exercise 4.2

Labeling. Write the name of the structure of each numbered part of the skin in the space provided. Color the different layers of the epidermis and dermis to highlight their structures.

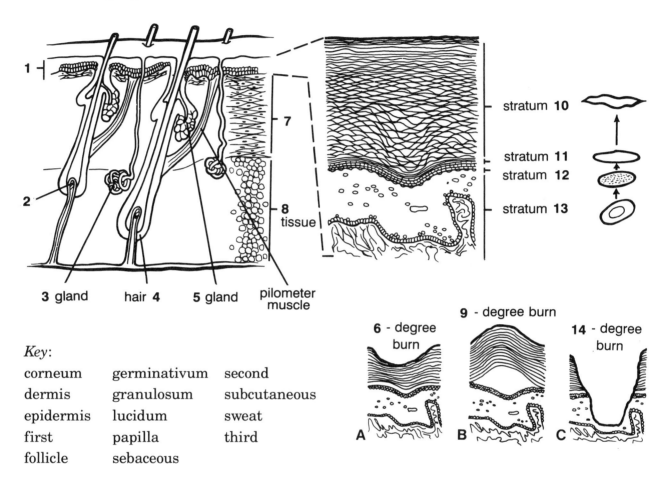

3 gland   hair **4**   **5** gland   pilometer
                                                     muscle

Key:

corneum	germinativum	second
dermis	granulosum	subcutaneous
epidermis	lucidum	sweat
first	papilla	third
follicle	sebaceous	

Figure 4.2 Diagram of a section of the skin to show its structure. The nucleated cell produced by the stratum germinativum dies (granulates) as it is forced outward to become the dead, scaly stratum corneum. The number of layers of the epidermis affected by the three types of skin burns is also shown. **A**. Only corneum cells are involved in first degree burns. **B**. Damage to the upper three layers occurs in second degree burns, forming a blister between layers 3 and 4. **C**. A third degree burn involves all epidermal layers and, therefore, usually requires a skin graft to replace the stratum germinativum.

1. _____   8. _____
2. _____   9. _____
3. _____   10. _____
4. _____   11. _____
5. _____   12. _____
6. _____   13. _____
7. _____   14. _____

V. TEST ITEMS

A. *Multiple Choice.* There is only one answer that is either correct or most appropriate. Circle the answer that corresponds to the question.

1. The basal layer of the epidermis rests upon which membranous tissue?
   a. basilar membrane
   b. perineal membrane
   c. basement membrane
   d. reticular membrane

2. The ridges that make up finger, foot, and palm prints have their origin in which skin layer?
   a. dermis
   b. epidermis
   c. subcutaneous tissue
   d. deep fascia

3. Which tissue layer is generously supplied with adipose or fat cells?
   a. dermis
   b. epidermis
   c. subcutaneous tissue
   d. deep fascia

4. Which of the following represents the outer nonvascular layer of the skin?
   a. epidermis
   b. dermis
   c. epithelium
   d. corium

5. The epidermis or outer skin layer is composed of which type of epithelial cells?
   a. ciliated columnar
   b. stratified squamous
   c. simple squamous
   d. simple columnar

6. Which cellular layer (stratum) of the epidermis is the only one capable of producing new cells?
   a. stratum corneum
   b. stratum lucidum
   c. stratum germinativum
   d. stratum granulosum

7. Which of the following constitutes a function of the skin?
   a. producing antibodies
   b. regulating body temperature
   c. integrating sensory activity
   d. facilitating the loss of body water

8. Select the highest level of organization from the following
   a. tissue
   b. system
   c. organ
   d. cell

9. Cilia are
   a. found in all epithelial structures.
   b. hairlike structures that move.
   c. hollow.
   d. secretory in function.

10. Connective tissues include
    a. bone.
    b. nerve.
    c. adipose.
    d. muscle.

11. A group of similar cells is called a (an)
    a. organ.
    b. organism.
    c. tissue.
    d. system.

12. Epithelium is a type of
    a. contracting tissue.
    b. lining tissue.
    c. secreting tissue.
    d. conducting tissue.

13. Complex genetic factors involving the varying composition of which substance determine the color of body skin?
    a. melanthin
    b. menthene
    c. merocrine
    d. melanin

14. Which glands located adjacent to hair follicles secrete an oily substance that keeps the hair and skin soft and pliable?
    a. sudoriferous glands
    b. sebaceous glands
    c. suprarenal glands
    d. serous glands

15. Connective tissue is best described as a
    a. supporting tissue.
    b. secreting tissue.

16. Simple unicellular glands that produce mucous are called
    a. tubular.
    b. goblet.
    c. acinous.
    d. alveolar.

17. All of these fibers would be found in connective tissue *except*
    a. reticular.
    b. elastic.
    c. spindle.
    d. collagenous.

18. Which of the following is not classified as a tissue?
    a. bone
    b. nerve
    c. blood
    d. skin

19. What is the function of ligaments?
    a. bind bones together at the joints
    b. cover the ends of bones
    c. connect muscle to bone
    d. support parts of cranial skeleton

20. Glands, such as the thyroid, that secrete their products directly into the blood are classified as
    a. exocrine.
    b. endocrine.
    c. sebaceous.
    d. ceruminous.

B. *Matching Questions.* Each of the phrases in COLUMN B refers to a word or phrase in COLUMN A. Insert the letter of the word or phrase from COLUMN B that best describes it. Some words may be used more than once or not at all.

Column A

1. ___ areolar connective tissue
2. ___ ciliated columnar epithelium
3. ___ fibrocartilage
4. ___ fibrous white connective tissue
5. ___ elastic cartilage
6. ___ hyaline cartilage
7. ___ irregular dense connective tissue
8. ___ transitional epithelium
9. ___ squamous epithelium

Column B

a. lining of urinary bladder
b. pleura
c. beneath the skin, between muscles, underneath epithelial cells
d. framework of ear and larynx
e. tendons and ligaments
f. ends of bones and tracheal rings
g. between vertebrae
h. lining of respiratory passageway
i. dermis of skin

Column A

1. ___ polyp
2. ___ melanoma
3. ___ albinism
4. ___ subcutaneous
5. ___ dermis

Column B

a. lack of skin pigment
b. under the skin
c. corium
d. mucous membrane tumor
e. corneum
f. malignant skin tumor

C. *True-False.* Place a *T* or *F* in the space provided.

___ 1. The color of skin is due to the presence of a pigment called keratin.

___ 2. Sweat glands of the skin are a type of simple coiled tubular gland.

___ 3. Goblet cells are single-celled glands found in certain connective tissues.

___ 4. The classification of epithelial tissues is based upon the shape and arrangement of cells.

___ 5. Epithelial tissue can be distinguished from connective tissue by sparseness of intercellular materials.

___ **6.** Cancer arises when a normal cell is transformed into a new growth.

___ **7.** Epithelial tissue has a rich blood supply.

___ **8.** A good example of elastic cartilage is found in the external ear.

___ **9.** All of the cells of the outer layer (stratum corneum) of the skin are dead.

___ **10.** A basement membrane seats or anchors epithelial tissue to underlying connective tissue.

Answer Sheet—Chapter 4

Exercise 4.1

EPITHELIUM

Epithelial tissue is a complex protective layer that <u>covers</u> the body and <u>lines</u> all the cavities and organs having a direct connection to the <u>outside</u> of the body. To put it simply, epithelium is a kind of <u>sheet</u> made up of a series of cells that <u>abut</u> each other. Epithelial cells are so closely joined together at these junctions that they form a <u>continuous</u> barrier between the body parts they cover and the surrounding medium (water, air, or internal body fluids). Epithelial tissue may be composed of only one flat layer of cells, or it may be <u>stratified</u> into different layers. When examined under a microscope, epithelial cells can be seen to have a supporting <u>basement</u> membrane, which appears in prepared sections as a fine line.

Types of Epithelial Tissue

Four distinct types of epithelium occur in the human body. Their classification is based on the <u>shapes</u> and properties of the cells composing them.

1. *Squamous epithelium.* Simple squamous epithelium is made up of flat cells arranged in a <u>single</u> layer. The term squamous means <u>scalelike</u>, and the cells of these tissues have <u>irregular</u> shapes like scales. Each cell contains a large, prominent nucleus at its center. In its simplest form, squamous epithelial tissue lines the small saclike structures of the lung (alveoli), which function in the <u>exchange</u> of <u>oxygen</u> and <u>carbon dioxide</u> during breathing. The lens of the eye is also made up of simple squamous epithelium.

2. *Cuboidal epithelium.* As its name suggests, cuboidal epithelium is made up of <u>cube-shaped</u> cells. This tissue lines the ducts of many <u>glands</u> and the <u>kidney</u>. The nuclei of its cells are spherical and usually are found in the center of the cell. Some cuboidal cells are capable of forming <u>secretions</u> and consequently are found in glands such as the thyroid, sweat glands, and salivary glands.

3. *Columnar epithelium.* Columnar epithelial cells are compressed to form <u>columnlike</u> shapes. These <u>tall</u> thin cells have a nucleus that usually can be found near the <u>basement</u> membrane. Epithelial tissues made up of these cells occur in the <u>digestive</u> tract, particularly in the intestines. Columnar epithelium is concerned primarily with the <u>secretion</u> of digestive fluids and with the <u>absorption</u> of food materials.

4. *Pseudostratified epithelium.* This fourth type of epithelium earned its name because on first glance it appears to be <u>arranged</u> in <u>layers</u>. This appearance is caused by variations in the <u>shape</u> of each cell making up the tissue. Although some of the cells in contact with the basement membrane do not reach the <u>surface</u> of the tissue, most of the cells are <u>tall</u> and do reach the surface. Pseudostratified epithelium is found most often in the <u>respiratory</u> tract, particularly in the <u>trachea</u> and in the <u>bronchi</u> of the lung.

Two modifications occurring in the cells of epithelial tissues are worth considering here: goblet cells and cilia. Goblet cells are flask shaped and contain mucoid secretion. Microscopically, they appear as open "empty" cells in their tissues. Cilia are hairlike appendages of the cell. With their continuous wavelike motion, they produce currents in the fluids at the cell's surface. Both goblet cells and cilia occur in columnar and pseudostratified epithelium. Goblet cells aid digestion by secreting mucus for absorption of partially digested foods; cilia aid in the passage of food down the intestinal tract. Goblet cells and cilia are also essential in the respiratory tract, where goblet cells add moisture to the air taken in and cilia clean the air of foreign particles that could otherwise clog the alveoli of the lungs.

Glands

Some epithelial tissues are made up of cells specifically organized to cause secretion.

Endocrine glands are sometimes called glands of internal secretion because they are situated far beneath the epithelial surface and have no ducts or passages by which their secretions can pass through the epithelium. Instead, they communicate directly with the circulatory system through the capillaries, which permit the distribution of their secretions throughout the body. The thyroid is an example of an endocrine gland.

Like the endocrine glands, exocrine glands are also located away from the epithelial surface. However, the exocrine glands are equipped with ducts that carry their secretions to the tissue surface. The salivary glands, for example, are exocrine glands. The pancreas is a gland that is both endocrine and exocrine. It produces digestive enzymes that are passed to the intestine through the pancreatic duct (exocrine); but also produces a hormone, insulin, which is transported through the body by the circulatory system (endocrine).

CONNECTIVE TISSUES

Connective tissues are found throughout the body. As their name implies, the principal function of these tissues is binding the body parts together. They form a framework for the internal organs; they also perform a variety of other functions, ranging from protection against injury to storage of fat.

A fundamental difference between connective tissue and epithelial tissue can be seen in their cellular compositions. Epithelial cells are directly adjacent to one another, separated only by a very small amount of intercellular substance called matrix. On the other hand, connective tissue contains few cells, and these are widely separated. The intercellular matrix is relatively abundant and usually determines the physical characteristics of a given connective tissue.

Connective Tissue Cells

In their embryonic stage, typical connective tissue cells are large and star shaped, with many projections called processes. These cells are fibroblasts, and they arise from an early embryonic tissue, the mesenchyme. This tissue develops into many different forms, including blood cells and muscle. As the connective tissues develop, the cells lose their star-shaped appearance and become widely separated by large amounts of intercellular material. The adult connective tissue cell, which is responsible for the formation of fibers, is the fibrocyte. Other types of cells found in connective tissue are:

1. *Histiocytes or macrophages.* These cells move about through the connective tissue, ingesting foreign materials, bacteria, and cellular debris (phagocytosis) .

2. *Plasma cells.* These small, irregular cells are associated with the formation of antibodies, an important part of the body's defense against foreign substances.

3. *Mast cells.* These are located near the blood vessels and are involved in the production of heparin, an anticoagulant. They are also important in the production of histamine in allergic reactions.

4. *Blood cells*. The white blood cells, such as lymphocytes, monocytes, and neutrophils, are often present in connective tissue, where their function is to <u>destroy</u> <u>bacteria</u> by phagocytosis.

5. *Fat cells*. These specialized cells <u>store</u> fats and oils. Microscopically, a fat cell resembles a <u>signet</u> ring because the stored fat pushes the nucleus and the cytoplasm to one side of the cell.

Connective Tissue Fibers

The <u>fibers</u> characteristic of connective tissue are found within the <u>intercellular</u> matrix. There are three general types:

1. *Collagenous fibers*. These white fibers contain <u>collagen</u>, an albumin that is the principal structural protein of the body. This protein is synthesized by fibrocytes and is the major constituent of skin, <u>ligaments</u>, <u>tendons</u>, <u>cartilage</u>, and <u>bone</u>. The fibers occur in bundles and are relatively nonelastic.

2 *Elastic fibers*. These fibers are <u>yellow</u> and may occur singly or in bundles. They are highly <u>elastic</u> and branch to unite with other fibers. Elastic fibers are both larger and straighter than collagenous fibers.

3. *Reticular fibers*. These short and very thin fibers branch freely, forming a <u>cobweb</u> network called a <u>reticulum</u>. These fibers are <u>nonelastic</u> and are usually found forming the internal framework of <u>glands</u>.

Types of Connective Tissue

Adipose tissue. <u>Areolar</u> tissue beneath the epidermis is commonly filled with <u>loosely</u> packed <u>fat</u> cells held together in bundles by collagenous and elastic fibers. <u>Adipose</u> (fat) tissue is found primarily as a kind of <u>padding</u> around the joints, as soft pads between the organs, around the kidney and the heart, and in the yellow marrow of long bones. Its fat cells provide a <u>reserve</u> food <u>supply</u> and also <u>insulate</u> the body against <u>heat</u> loss.

Dense connective tissue. Many parts of the body require a <u>stronger</u> type of connective tissue to <u>enclose</u>, <u>restrain</u> or <u>separate</u> functioning structures. The basic forms of dense connective tissue are <u>tendons</u>, which attach <u>muscle</u> to bone; <u>ligaments</u>, which connect the <u>bones</u> that form joints; <u>aponeuroses</u>, which are thin, tendinous sheets attached to flat <u>muscles</u>; and <u>fasciae</u>, the thin sheets of tissue that <u>cover</u> muscles and hold them in place.

Cartilage. Cartilage is a <u>modified</u> form of connective tissue in which cartilage cells (<u>chondrocytes</u>) lie in small <u>capsules</u>, the <u>lacunae</u>, which are surrounded by an irregular matrix made up of fibers and/or a gel.

<u>Hyaline</u> cartilage, the most common form of cartilage, is a modified areolar tissue with a white or glassy appearance. It may contain any number of unevenly dispersed lacunae. Hyaline cartilage forms the <u>skeleton</u> in the embryo and <u>covers</u> the articulating surfaces of bones in the joint cavities. It also provides the structure for the nose and the connections of the ribs to the breast bone, and, in the respiratory tract, forms the ringlike trachea and bronchi.

<u>Elastic</u> cartilage contains many elastic fibers and is found wherever cartilage is required to <u>move</u>, such as in the epiglottis and the external ear.

<u>Fibrous</u> cartilage is less <u>rigid</u> than hyaline, but contains heavy bundles of collagenous connective tissue. It is found in the <u>intervertebral</u> disks, which absorb shocks between the vertebrae of the backbone. Fibrous cartilage also <u>reinforces</u> the hyaline articular cartilages at the knee and hip.

Exercise 4.2

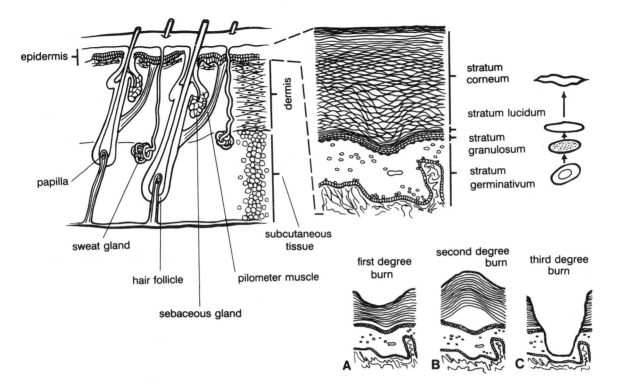

Figure 4.2   Diagram of a section of the skin to show its structure. The nucleated cell produced by the stratum germinativum dies (granulates) as it is forced outward to become the dead, scaly stratum corneum. The number of layers of the epidermis affected by the three types of skin burns is also shown. **A**. Only corneum cells are involved in first degree burns. **B**. Damage to the upper three layers occurs in second degree burns, forming a blister between layers 3 and 4. **C**. A third degree burn involves all epidermal layers and, therefore, usually requires a skin graft to replace the stratum germinativum.

Test Items

A.   1.c, 2.a, 3.c, 4.a, 5.b, 6.c, 7.b, 8.b, 9.b, 10.c, 11.c, 12.b, 13.d, 14.b, 15.a, 16.b, 17.c, 18.d, 19.a, 20.b.

B.   1.c, 2.h, 3.g, 4.e, 5.d, 6.f, 7.i, 8.a, 9.b.
     1.d, 2.f, 3.a, 4.b, 5.c.

C.   1.F, 2.T, 3.F, 4.T, 5.T, 6.T, 7.F, 8.T, 9.T, 10.T.

Tissues

Across

2 a malignant growth of epitelial origin

4 not a threat to life

5 a secreting organ

6 the true skin

9 a ductless gland

11 tough fibrous protein

12 outer layer of the epidermis

14 the outer layer of skin

19 a new growth or tumor

Down

1 a protein of skin and connective tissue

2 a malignant growth or condition

3 supporting tissue

7 a covering, the skin

8 a death threatening situation

9 tissue that covers the body and lines all cavities connected to the outside

10 a hollow cavity or capsule

13 a group of cells of similar origin and function

15 a gland with ducts

16 dark pigment of the skin

17 a malignant tumor of connective tissue origin

18 a small projection or elevation

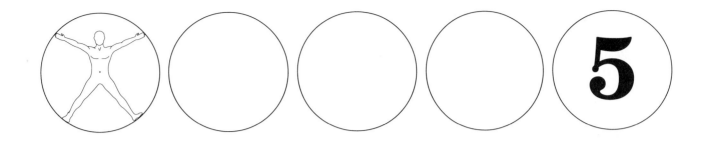

The Skeletal System

I. CHAPTER SYNOPSIS

The chapter considers the major functions of the skeletal system, the histology of bone, intramembranous and endochondral ossification, and bone growth and destruction as an example of the homeostasis of the bone.

The student is introduced to the four principal kinds of bones, important bone markings, and the bones of the axial and appendicular skeletons.

II. OBJECTIVES

After reading the chapter, the student should be able to:

- List the functions of the skeletal system.

- Describe the microscopic structure of the bone.

- Identify the gross anatomy of a bone.

- Distinguish between the axial skeleton and the appendicular skeleton, and name the components of each.

- Identify the bones that make up the pectoral and pelvic girdle.

III. IMPORTANT TERMS

Using your textbook, define the following terms:

appendicular (ap-en-dik'-yah-lur) _____

_____

atlas (at'-lus) _____

_____

axial (ak'-see-al) _____

_____

axis (ack'-sis) _____

_____

calcification (kal-sah-fah-kay'-shun) _____

_____

canal (kah-nal') _____

_____

cancellous (kan'-se-lus) _____

_____

cervical (ser'-vi-kal) _____

_____

compact (kom'-packt) _____

_____

diaphysis (di-af'-ah-sis) _____

_____

epiphysis (e-pif'-ah-sis) _____

_____

foramen (fo-ray'-men) _____

_____

fossa (fas'-ah) _____

_____

fracture (frak'-shur) _____

_____

lumbar (lum'-bar) _____

_____

ossification (os-i-fi-kay'-shun) _____

_____

osteocyte (os'-tee-o-sight) _____

_____

periosteum (per-ee-os'-tee-um) _____

_____

process (pros'-es) _____

_____

sesamoid (ses'-ah-moid) _____

_____

spine (spine) _____

_____

suture (soo'-cher) _____

_____

thoracic (tho-ras'-ik) _____

_____

trabeculae (trah-bek'-u-lee) _____

_____

vertebra (vurt'-ah-brah) _____

_____

IV. EXERCISES

Complete the following exercises in the order given. A precise set of terms and diagrams has been chosen to describe the skeletal system.

Exercise 5.1

Labeling.   Write the name of the bone in the space provided. Color the axial skeleton different from the appendicular skeleton.

Key:

carpal	metacarpal	scapula
clavicle	metatarsals	skull
femur	patella	sternum
fibula	phalanges	tarsals
humerus	pubis	tibia
ilium	radius	ulna
ischium	ribs	vertebrae

1. _____
2. _____
3. _____
4. _____
5. _____
6. _____
7. _____
8. _____
9. _____
10. _____
11. _____
12. _____
13. _____
14. _____
15. _____
16. _____
17. _____
18. _____
19. _____
20. _____
21. _____

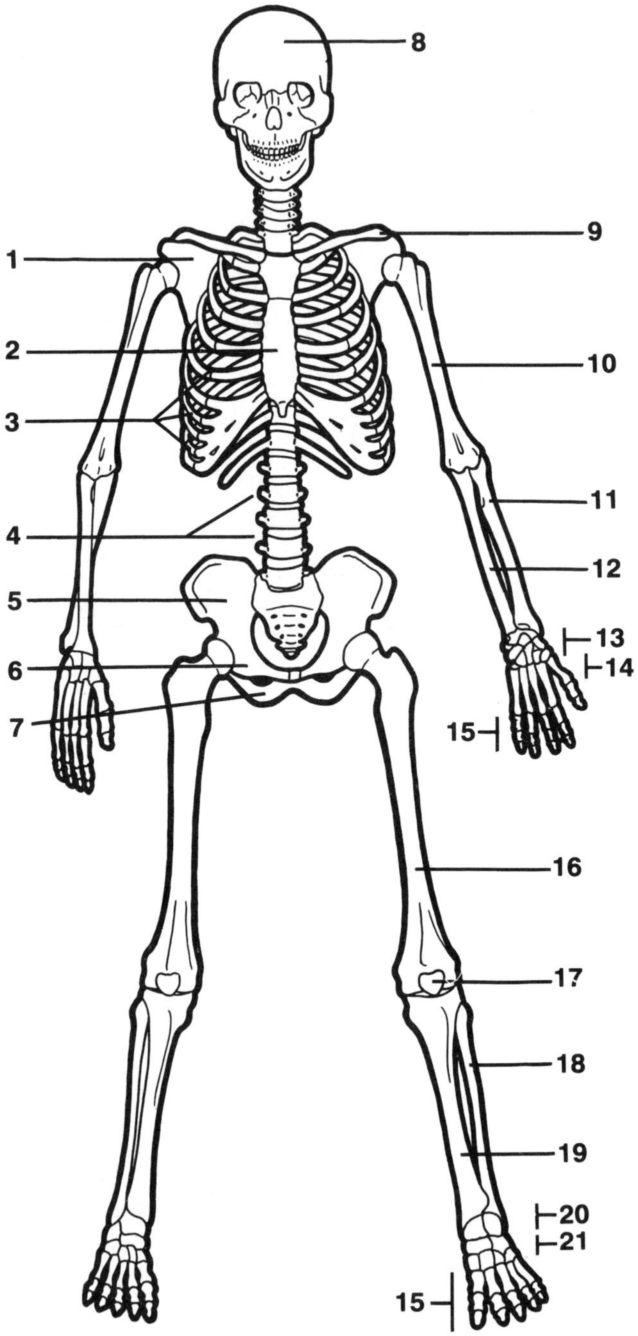

Figure 5.1   Human skeleton, anterior aspect.

Exercise 5.2

Labeling. Write the name of the structure of the bone in the space provided. Color the total bone different from the canal network.

Key:
blood vessel
cancellous bone
compact bone
diaphysis
epiphysis
epiphyseal line
Haversian canal
Haversian system
marrow
medullary cavity
periosteum
trabeculae
Volkmann's canal

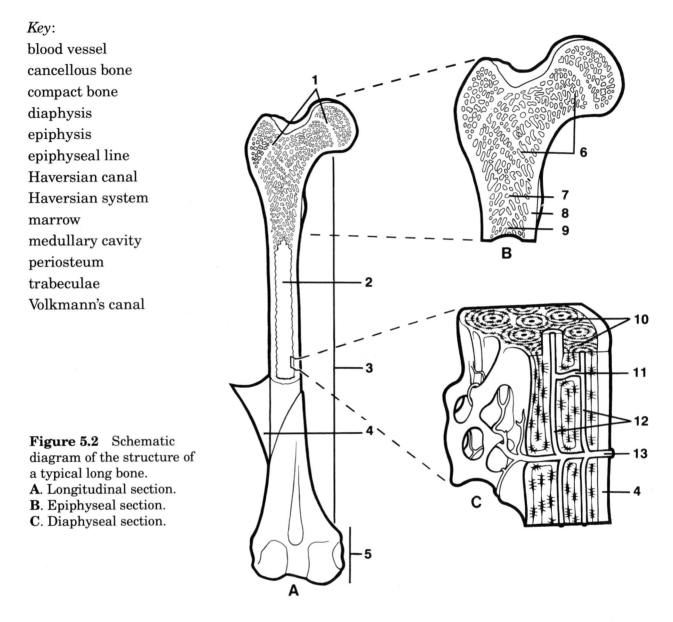

Figure 5.2 Schematic diagram of the structure of a typical long bone.
A. Longitudinal section.
B. Epiphyseal section.
C. Diaphyseal section.

1. _____
2. _____
3. _____
4. _____
5. _____
6. _____
7. _____

8. _____
9. _____
10. _____
11. _____
12. _____
13. _____

Exercise 5.3

Labeling. Write the name of the bone and its structures in the space provided. Color each bone differently.

Key:
coronal suture
ethmoid
frontal
infraorbital foramen
mandible
mastoid process
maxillae
mental foramen
nasal
nasal conchae
optic foramen
parietal
sphenoid
squamous suture
temporal
vomer
zygomatic

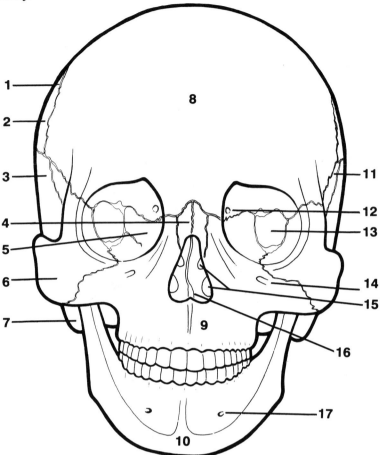

Figure 5.3 Frontal, lacrimal, zygomatic, maxillary, and nasal bones of the skull.

1. _____
2. _____
3. _____
4. _____
5. _____
6. _____
7. _____
8. _____
9. _____

10. _____
11. _____
12. _____
13. _____
14. _____
15. _____
16. _____
17. _____

Exercise 5.4

Labeling. Write the name of the bone and its structures in the space provided. Color code this exercise with the frontal view (Exercise 5.3).

Key:
acoustic meatus
coronal suture
ethmoid
frontal
lacrimal
lacrimal fossa
lambdoid suture
mandible
mastoid process
maxilla
mental foramen
nasal
occipital
parietal
sphenoid
squamous suture
styloid process
temporal
zygomatic arch
zygomatic bone

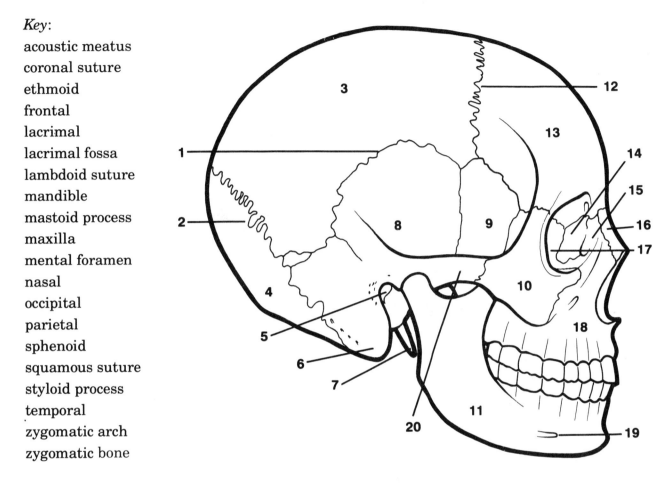

Figure 5.4   Skull, lateral aspect.

1. _____
2. _____
3. _____
4. _____
5. _____
6. _____
7. _____
8. _____
9. _____
10. _____

11. _____
12. _____
13. _____
14. _____
15. _____
16. _____
17. _____
18. _____
19. _____
20. _____

V. TEST ITEMS

A. *Multiple Choice.*   There is only one answer that is either correct or most appropriate. Circle the answer that corresponds to the question.

1. Bone formation is referred to as
   a. osteomyelitis.
   b. osteoporosis.
   c. rickets.
   d. ossification.

2. Which one of the following is a bone reabsorbing or destroying cell?
   a. osteocyte
   b. osteoclast
   c. osteoblast
   d. bone cell

3. Which one of the following is *not* included in the thoracic cage?
   a. ribs
   b. thoracic vertebrae
   c. clavicle
   d. sternum

4. The human skeleton functions in
   a. support.
   b. protection.
   c. muscle attachment.
   d. all of these

5. A person with a fractured patella has sustained a break in which type of bone?
   a. short
   b. long
   c. sesamoid
   d. sutural

6. A vitamin D deficiency in an adult results in a condition called
   a. osteosarcoma.
   b. osteomyelitis.
   c. osteoma.
   d. osteomalacia.

7. The death of osseous tissue from the deprivation of blood supply is called
   a. osteomyelitis.
   b. osteoarthritis.
   c. necrosis.
   d. osteoblastoma.

8. The epiphyses of long bones are covered by
   a. periosteum.
   b. endosteum.
   c. Volkmann's canals.
   d. articular cartilage.

9. Concentric rings of calcified intercellular substance arranged around Haversian canals are called
   a. lamellae.
   b. lacunae.
   c. trabeculae.
   d. canaliculi.

10. If you are told that a patient has a Pott's fracture, you know immediately that the bone involved is the
    a. humerus.
    b. radius.
    c. femur.
    d. fibula.

11. The improper fusion of portions of which bone result in cleft palate?
    a. maxilla
    b. sphenoid
    c. ethmoid
    d. mandible

12. A patient with a lateral curvature of the spine to the left would have which of the following conditions?
    a. scoliosis
    b. kyphosis
    c. lordosis
    d. hunchback

13. Severe pain behind the external auditory meatus most likely involves the
    a. sphenoid bone.
    b. mastoid process.
    c. parietal bone.
    d. zygomatic arch.

14. An individual exhibits the following symptoms: degeneration of epiphyseal cartilage, poor calcification, bow legs, and malformations of the head and pelvic bones. It is likely that the individual is suffering from
    a. osteoporosis.
    b. rickets.
    c. osteomyelitis.
    d. bone cancer.

15. The remodeling of bone is a function of which cells?
    a. chondrocytes and osteocytes
    b. osteoblasts and osteocytes
    c. osteocytes and osteoclasts
    d. chondroblasts and osteoclasts

16. A disorder of bones closely related to decreased activity of osteoblasts due to a hormone deficiency is
    a. osteomyelitis.
    b. osteoporosis.
    c. osteosarcoma.
    d. osteomalacia.

17. The growth of a long bone in length occurs at the
    a. epiphyseal plate.
    b. articular cartilage.
    c. epiphyseal line.
    d. center of the shaft.

18. A partial fracture in which one side of the bone is broken and the other side bends is called
    a. comminuted.
    b. transverse.
    c. spiral.
    d. greenstick.

19. A Colles' fracture involves the
    a. ulna.
    b. tibia.
    c. fibula.
    d. radius.

20. A compound fracture means
    a. the bone is fractured in several places.
    b. the bone is splintered into several small fragments.
    c. the broken end or ends of the bone protrude through the skin.
    d. the broken ends of the bone are driven into each other.

B. *Matching Questions.* Each of the phrases in COLUMN B refers to a word or phrase in COLUMN A. Insert the letter of the word or phrase from COLUMN B that best describes it. Some words may be used more than once or not at all.

Column A		*Column B*

    1. ___ fossa

    2. ___ tubercle

    3. ___ spine

    4. ___ foramen

    5. ___ head

    6. ___ crest

    7. ___ condyle

    8. ___ paranasal sinus

    9. ___ meatus

10. ___ tuberosity

11. ___ groove

12. ___ trochanter

a. a rounded opening through which blood vessels, nerves, and ligaments pass

b. a prominent border or ridge on a bone

c. an air-filled cavity within a bone connected to the nasal cavity

d. a small, rounded process

e. a large, rounded, usually roughened process

f. a rounded projection supported on a neck

g. a sharp, slender process

h. a depression in or on a bone

i. a relatively large, convex knucklelike prominence

j. a very large, blunt projection found only on the femur

k. a tubelike passageway running within a bone

l. a furrow that accommodates soft structures such as a blood vessel, nerve, or tendon

C. *True-False.* Place a *T* or *F* in the space provided.

___ **1.** The flat bones of the skull develop via intramembranous ossification.

___ **2.** Lamellae are found only in cancellous bone.

___ **3.** The small passages connecting the lacunae are termed Haversian canals.

___ **4.** The spaces in cancellous bone are filled with red bone marrow.

___ **5.** Red bone marrow manufactures both white and red cells.

___ **6.** The greatest concentration of calcium in the body is in the blood.

___ **7.** Long bones are larger versions of short bones.

___ **8.** The humerus is an example of a long bone.

___ **9.** The styloid process is part of the ethmoid bone.

___ **10.** The three types of moveable vertebrae are cervical, thoracic, and lumbar.

___ **11.** The axis is the second thoracic vertebra.

___ **12.** Hollowing of the bone is brought about by the action of osteoclasts.

___ **13.** In intramembranous bone formation, a "scale model" of hyaline cartilage is replaced by bone.

___ **14.** The sagittal suture joins the parietal bones of the skull.

___ **15.** The parietal bone articulates with the vertebral column.

___ **16.** The last five ribs are called false ribs.

___ **17.** The bony portion of the nasal septum is formed principally by the vomer and ethmoid bones.

___ **18.** The temporal bone contains the middle and inner ear.

___ **19.** The type of cartilage that serves as a model for the structure of all cartilage is elastic cartilage.

___ **20.** A periosteum surrounds bone.

Answer Sheet—Chapter 5

Exercise 5.1

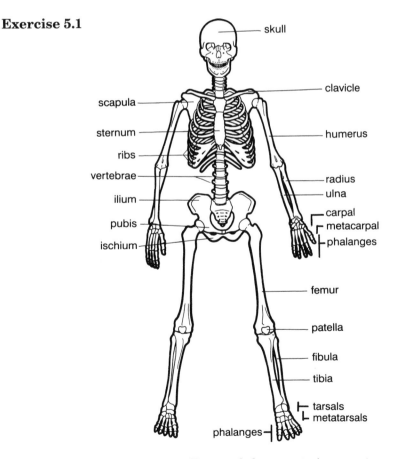

Figure 5.1 Human skeleton, anterior aspect.

Exercise 5.2

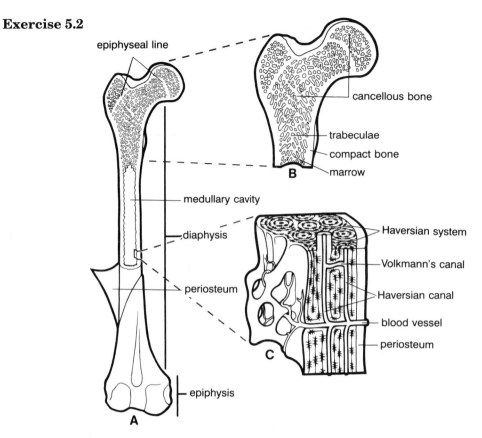

Figure 5.2   Schematic diagram of the structure of a typical long bone. **A**. Longitudinal section. **B**. Epiphyseal section. **C**. Diaphyseal section.

Exercise 5.3

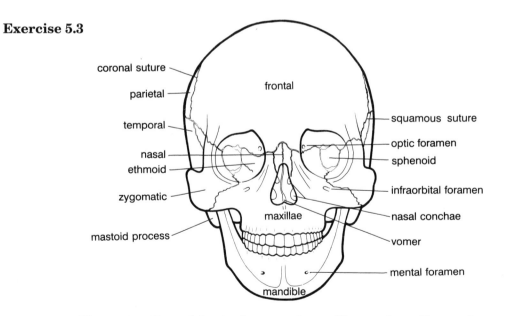

Figure 5.3   Frontal, lacrimal, zygomatic, maxillary, and nasal bones of the skull.

Exercise 5.4

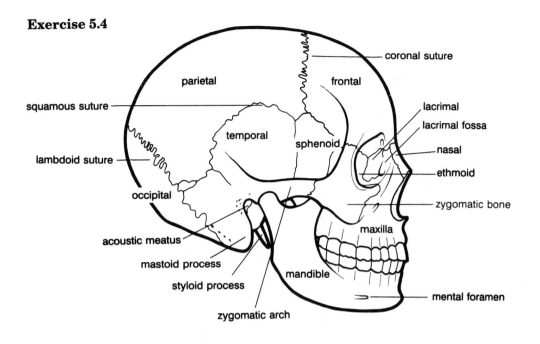

Figure 5.4 Skull, lateral aspect.

Test Items

A. 1.d, 2.b, 3.c, 4.d, 5.c, 6.d, 7.c, 8.d, 9.a, 10.d, 11.a, 12.a, 13.b, 14.b, 15.c, 16.b, 17.a, 18.d, 19.d, 20.c.

B. 1.h, 2.d, 3.g, 4.a, 5.f, 6.b, 7.i, 8.c, 9.k, 10.e, 11.l, 12.j.

C. 1.T, 2.F, 3.F, 4.T, 5.T, 6.F, 7.F, 8.T, 9.F, 10.T, 11.F, 12.T, 13.F, 14.T, 15.F, 16.T, 17.T, 18.T, 19.F, 20.T.

Skeletal System

Across

2 a bone that forms the spinal cord

4 that part of the skeleton that hangs from the main frame

10 to harden into bone

12 a narrow tubular passage

13 pertaining to the neck

15 seed-like

17 pertaining to the lower back

18 natural opening or hole

19 a broken bone

Down

1 connective tissue layer covering the bone

3 a mature bone cell

4 the first cervical vertebra

5 the end of a bone

6 the shaft of a bone

7 to harden

8 the second cervical vertebra

9 a prominence or projection

11 spongy bone

14 hard bone

15 thorn-like process of projection

16 junction between two immovable bones

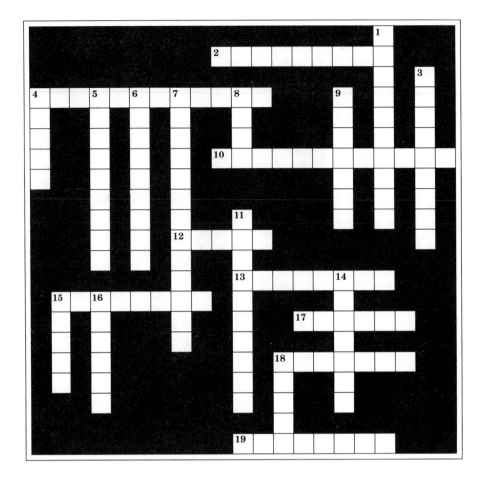

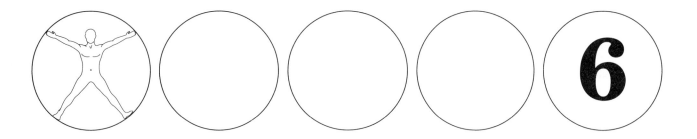

The Articular System

I. CHAPTER SYNOPSIS

The student is introduced to the various kinds of joints in the body. Articulations or joints can be divided into three classes: synarthroses, or immovable joints; amphiarthroses, or slightly movable joints; and diarthroses, or freely movable joints. The six subgroups of the diarthrosis class are described along with the four types of angular movements and seven types of rotational movements that are possible at diarthrotic joints.

The most difficult part of the chapter deals with the planes of movement and axes of rotation. Some students grasp the idea almost intuitively, while others never seem to really understand it. For those who do, it makes muscle actions much easier to understand (and for instructors to explain) and reduces the amount of memorizing required.

II. OBJECTIVES

After reading the chapter, the student should be able to:

- Classify body joints.
- Describe the distinguishing features of a synovial joint.
- Name the types of synovial joints and give an example of each.
- Describe the movements at synovial joints and give examples.

- Define arthritis and describe the different types.
- Define bursitis and relate, anatomically, its importance to joints.

III. IMPORTANT TERMS

Using your textbook, define the following terms:

abduction (ab-duk'-shun) _____

_____

adduction (ah-duk'-shun) _____

_____

amphiarthroses (am-fee-ahr-thro'-sis) _____

_____

ankylosis (an-ki-lo'-sis) _____

_____

arthritis (ar-thright'-us) _____

_____

articulate (ar-tik'-yoo-lut) _____

_____

articulation (ar-tik-yoo-lay'-shun) _____

_____

bursitis (bur-sight'-us) _____

_____

circumduction (ser-kum-duk'-shun) _____

_____

diarthroses (di-ar-thro'-sis) _____

_____

dislocation (dis-lo-kay'-shun) _____

_____

eversion (ee-ver'-zhun) _____

_____

extension (ek-sten'-shun) _____

_____

fibrositis (fibe-ro-sight'-us) _____

_____

flexion (flek'-shun) _____

_____

inversion (in-ver'-zhun) _____

_____

joint (joint) _____

_____

ligament (lig'-ah-ment) _____

_____

pronation (pro-nay'-shun) _____

_____

protraction (pro-trak'-shun) _____

_____

retraction (re-trak'-shun) _____

_____

rotation (ro-tay'-shun) _____

_____

sprain (sprane) _____

_____

supination (soo-pah-nay'-shun) _____

_____

symphysis (sim'-fi-sis) _____

_____

synarthroses (sin-ar-thro'-sis) _____

_____

synovia (sin-no'-vee-ah) _____

_____

synovial (sin-no'-vee-al) _____

_____

IV. EXERCISES

Complete the following exercises in the order given. A precise set of terms
and diagrams has been chosen to describe the articular system.

Exercise 6.1

Labeling.   Write the name of the type of joint or structure in the space
provided. Color each joint differently.

Key:
amphiarthrodial
cartilage
fibrocartilage
fibrous
intervertebral
pubic
suture
synarthrodial

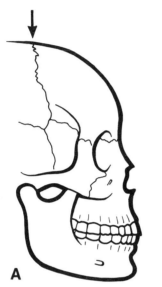

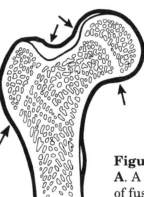

Figure 6.1 —1— joints.
A. A —2— (arrow) composed
of fusing —3— tissue. **B.** A
femoral head held in position by fusing —4— (syn-
chondrosis) (arrows).
—5— joints. **C.** Diagram of
an —6— disk (arrows),
lateral aspect. **D.** The
—7— bones are joined by a
disk of —8— forming the
symphysis pubis (arrows).

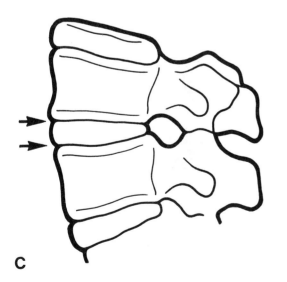

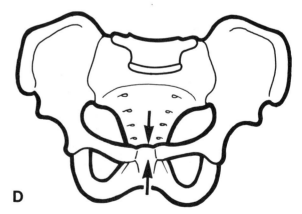

1. _____

2. _____

3. _____

4. _____

5. _____

6. _____

7. _____

8. _____

Exercise 6.2

Labeling.   Identify the types of diarthrodial joints by placing the answer in the space provided. Color each joint separately.

Key:
ball-and-socket
ellipsoidal
gliding
hinge
pivot
saddle

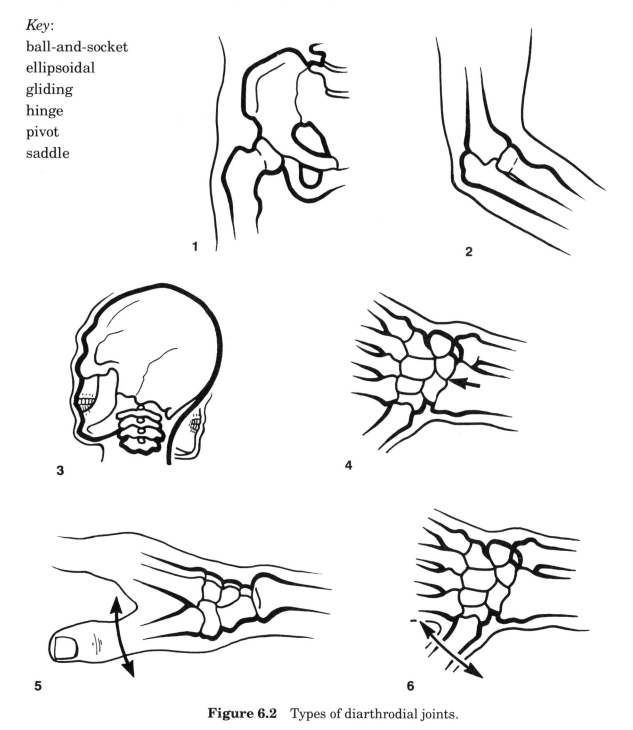

Figure 6.2   Types of diarthrodial joints.

1. _____   4. _____

2. _____   5. _____

3. _____   6. _____

Exercise 6.3

Labeling. Write the name of the movement at each joint in the space provided. Color the arrows to indicate the motion.

Key:
abduction-adduction
circumduction
eversion-inversion
flexion-extension
rotation
protraction-retraction
supination-pronation

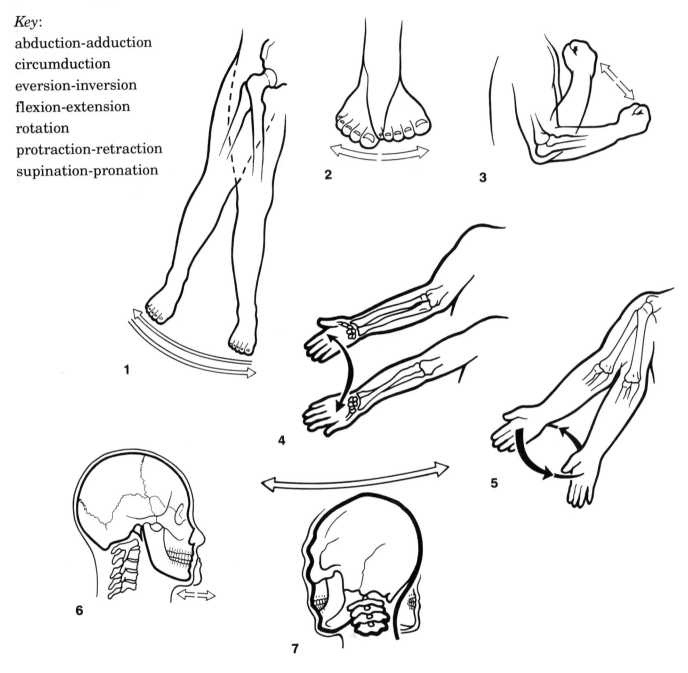

Figure 6.3 Movements of diarthrodial joints.

1. _____ 5. _____
2. _____ 6. _____
3. _____ 7. _____
4. _____

Exercise 6.4

Labeling. Write the name of the structure of the true joint in the space provided. Color the coronal section (B).

Key:

articular	joint	synovial
femur	ligament	tibia
fibula	patella	

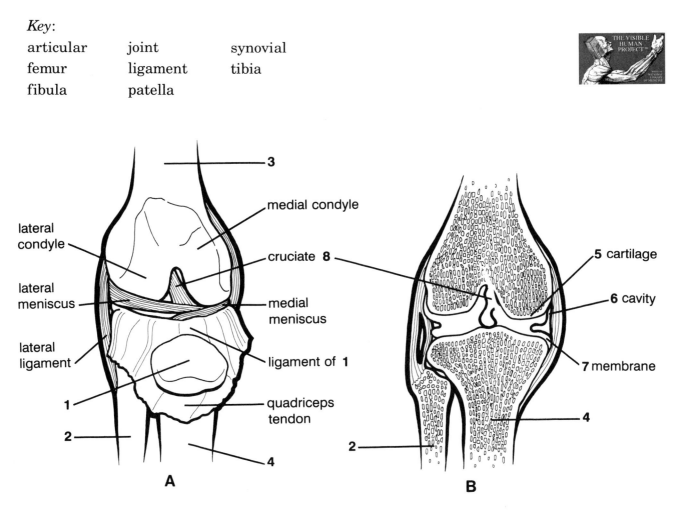

A

B

labels in A: 3, medial condyle, lateral condyle, lateral meniscus, lateral ligament, cruciate **8**, medial meniscus, ligament of **1**, quadriceps tendon, **1**, **2**, **4**

labels in B: **5** cartilage, **6** cavity, **7** membrane, **4**, **2**

Figure 6.4   Right knee joint. **A.** Dissected from the front, with the patella resected and hanging down onto the tibia. **B.** Coronal section of the knee, exposing the joint cavity.

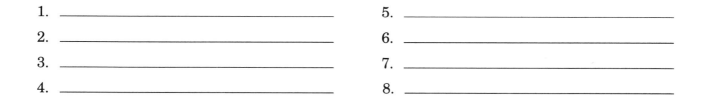

1. _____   5. _____

2. _____   6. _____

3. _____   7. _____

4. _____   8. _____

V. TEST ITEMS

A. *Multiple Choice.* There is only one answer that is either correct or most appropriate. Circle the answer that corresponds to the question.

1. The term used to describe overstretching or pulling a part of a muscle is
   a. sprain.   c. strain.
   b. dislocation.

2. The term for displacement of a bone from its normal position in a joint is
   a. sprain.   c. strain.
   b. dislocation.

3. Which movements can you perform at both your hip and knee joints?
   a. abduction and extension
   b. abduction and adduction
   c. circumduction and rotation
   d. flexion and extension

4. The diarthrodial joint in which motion is limited to rotation is termed
   a. hinge.   c. pivot.
   b. flexor.   d. condyloid.

5. When the sole of the foot is turned toward the midline of the body it is said to be
   a. extended.   c. inverted.
   b. pronated.   d. everted.

6. Bursae are small connective sacs lined with synovial membrane and containing
   a. fat-filled folds of tissue.
   b. viscous fluid.
   c. areolar tissue.
   d. hyaline cartilage.

7. The hip joint and shoulder joint are examples of which type of joint?
   a. hinge   c. ball-and-socket
   b. pivot   d. plane

8. In the disorder referred to as torn cartilage, which portion of the joint is injured?
   a. articular cartilage
   b. meniscus
   c. synovial membrane
   d. fibrocartilage

9. A joint that affords three planes of movement is the
   a. saddle.   c. ball-and-socket.
   b. ellipsoidal.   d. hinge.

10. A movement in which the distal end of a bone moves in a circle while the proximal end remains relatively stable is called
   a. rotation.   c. protraction.
   b. circumduction.   d. supination.

11. Which of the following is *not* a synovial joint?
    a. symphysis   c. gliding
    b. pivot   d. ball-and-socket

12. Classification of joints is based on
    a. location in the body.
    b. presence or absence of tendons.
    c. movability.
    d. length of bones concerned.

13. The clinical term applied to inflammation of a tendon and synovial membrane at a joint and commonly referred to as tennis elbow is
    a. gout.   c. tendinitis.
    b. osteoitis.   d. bursitis.

14. The statement "to draw away laterally from the median plane of the body" best describes what word?
    a. flexion   d. pronation
    b. adduction   e. rotation
    c. abduction

15. Extension of the foot at the ankle joint is known as
    a. hyperextension.   c. dorsiflexion.
    b. plantar flexion.   d. abduction.

16. The statement "stretching out" best defines what word?
    a. extension   d. supination
    b. flexion   e. pronation
    c. abduction

17. The statement "to move backward" best describes which of the following words?
    a. supination   d. eversion
    b. adduction   e. retraction
    c. abduction

18. Inversion is defined as
    a. moving backward.
    b. the upward position of the palm of the hand.
    c. abduction of the foot.
    d. adduction of the foot.
    e. to make an angle.

19. A joint of the body that contains a broad, flat disk of fibrocartilage would be classified as
    a. ball-and-socket joint.
    b. suture.
    c. symphysis joint.
    d. gliding joint.

20. Lying on one's stomach is termed
    a. supination
    b. pronation.
    c. protraction.
    d. retraction.

B. *Matching Questions.* Each of the phrases in COLUMN B refers to a word or phrase in COLUMN A. Insert the letter of the word or phrase from COLUMN B that best describes it. Some words may be used more than once or not at all.

Column A	*Column B*
1. ___ bursectomy	**a.** inflammation of a synovial membrane of a joint
2. ___ arthralgia	**b.** displacement of a bone from its natural position in a joint
3. ___ synovitis	**c.** removal of a bursa
4. ___ chondritis	**d.** inflammation of a cartilage
5. ___ arthrosis	**e.** severe or complete loss of movement at a joint
6. ___ dislocation	**f.** pain in a joint
7. ___ ankylosis	**g.** tearing of tendons and ligaments
8. ___ sprain	**h.** disease of a joint

Column A	*Column B*
1. ___ bunion	**a.** cone forming
2. ___ protraction	**b.** gouty arthritis
3. ___ tenosynovitis	**c.** toward the midline
4. ___ rotation	**d.** rheumatoid arthritis
5. ___ fibrous ankylosis	**e.** body part forward
6. ___ ureate crystals	**f.** palm down
7. ___ adduction	**g.** foot turns in
8. ___ pronation	**h.** inflamed tendon sheath
9. ___ eversion	**i.** foot turns out
10. ___ circumduction	**j.** palm forward
	k. callus
	l. around central axis

C. *True-False.* Place a *T* or *F* in the space provided.

___ 1. Synarthroses do not permit movement.

___ 2. The synovial membrane is found only in diarthrodial joints and bursae.

___ 3. Joint stability is provided by synovial fluids.

___ 4. The saddle joint for the thumb refers to the joint between a metacarpal and a phalangeal bone.

___ 5. Supination and pronation of the lower arm and hand occur when the radius rotates around the ulna.

___ **6.** All diarthroses permit free movement but not necessarily the same kinds of movements between articulating bones.

___ **7.** All diarthroses have a joint capsule and a joint cavity.

___ **8.** The term synarthroses is another name for synovial joints.

___ **9.** Most joints in the body are synovial.

___ **10.** Most diarthroses are ball-and-socket joints.

___ **11.** Both the knee joint and the elbow joint are classified as hinge synovial joints.

___ **12.** Cartilaginous joints permit no movement between the articulating bones.

___ **13.** No diarthroses permits all of the following movements: flexion, extension, abduction, adduction, rotation, and circumduction.

___ **14.** Flexions are bending movements, whereas extensions are straightening movements.

___ **15.** Moving the forearm so as to turn the palm forward, as it is in the anatomical position, is called supination.

Answer Sheet—Chapter 6

Exercise 6.1

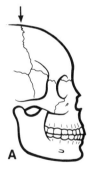
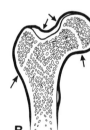

Figure 6.1 Synarthrodial joints. **A.** A suture (arrow) composed of fusing fibrous tissue. **B.** A femoral head held in position by fusing cartilage (synchondrosis) (arrows). Amphiarthrodial joints. **C.** Diagram of an intervertebral disk (arrows), lateral aspect. **D.** The pubic bones are joined by a disk of fibrocartilage forming the symphysis pubis (arrows).

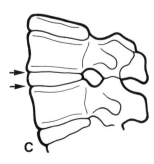
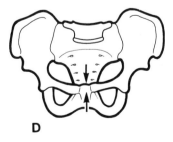

Exercise 6.2

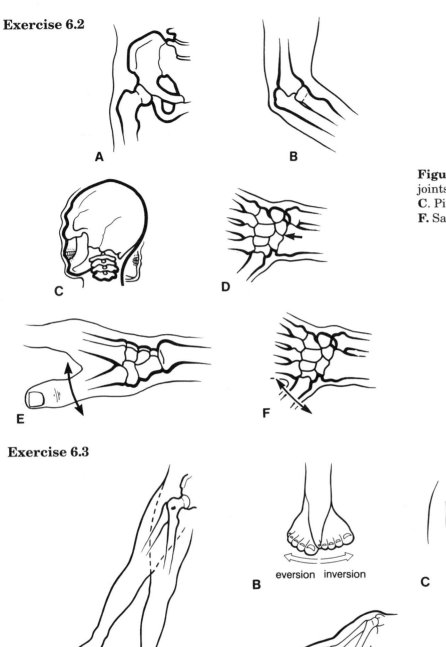

A

B

C

D

E

F

Figure 6.2. Types of diarthrodial joints. **A**. Ball-and-socket. **B**. Hinge. **C**. Pivot. **D**. Ellipsoidal. **E**. Gliding. **F**. Saddle.

Exercise 6.3

A

abduction
adduction

B

eversion   inversion

C

flexion

extension

D

circumduction

E

supination

pronation

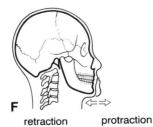

F

retraction   protraction

G

rotation

Figure 6.3   Movements of diar-throdial joints. **A**. Abduction-adduction. **B**. Eversion-inversion. **C**. Flexion-extension. **D**. Circum-duction. **E**. Supination-prona-tion. **F**. Protraction-retraction. **G**. Rotation.

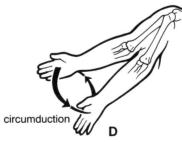
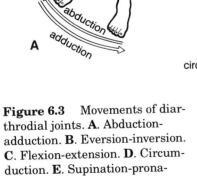
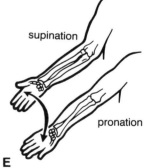

Exercise 6.4

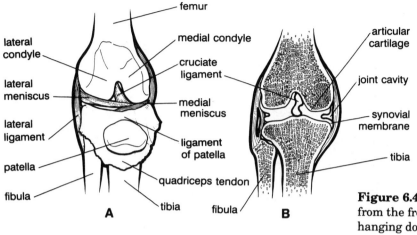

Figure 6.4. Right knee joint. **A**. Dissected from the front, with the patella resected and hanging down onto the tibia. **B**. Coronal section of the knee, exposing the joint cavity.

Test Items

A. 1.c, 2.b, 3.d, 4.c, 5.c, 6.b, 7.c, 8.b, 9.c, 10.b, 11.a, 12.c, 13.c, 14.c, 15.b, 16.c, 17.e, 18.d, 19.c, 20.b.

B. 1.c, 2.f, 3.a, 4.d, 5.h, 6.b, 7.e, 8.g.
   1.k, 2.e, 3.h, 4.l, 5.d, 6.b, 7.c, 8.f, 9.i, 10.a.

C. 1.T, 2.T, 3.F, 4.F, 5.T, 6.T, 7.T, 8.F, 9.T, 10.F, 11.T, 12.F, 13.F, 14.T, 15.T.

Chapter 6

Articular System

Across

1 membrane of a true joint

7 inflammation of a joint

13 circular movement of a part

16 displacement of a member from its joint

17 bending a limb or body part

Down

1 a joint injury

2 turning a body part inward

3 connective tissue connecting bone to bone

4 straightening of a body part

5 point of connection between two bones

6 a freely moving joint

8 an immovable joint

9 to move between bones

10 lying face down or crossing your lower arms

11 inflammation of a bursa sac

12 fusion of a joint

14 movement toward the midline

15 movement around the central axis

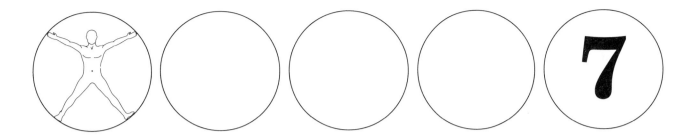

The Muscles

I. CHAPTER SYNOPSIS

The end result of a control system is the activation of an effector organ that can produce a change in the environment. The two major effector organs in the body are muscles and glands.

Muscle tissue, owing to its ability to contract or shorten, and thus produce movement of internal and external body parts, is responsible for such basic life processes as heartbeat, respiration, digestion, and production of body heat as well as for gross skeletal movements. There are three distinct types of muscle tissue: skeletal, cardiac, and smooth. Each type has a uniquely characteristic microscopic structure, gross structure, nerve supply, and blood supply, and each is designed for the performance of a certain type of work. Although all three types of muscle tissue have the same fundamental properties of excitability and contractility, certain differences are also present.

II. OBJECTIVES

After reading the chapter, the student should be able to:

- Describe the structure and control of three different types of muscles.

- Describe the microscopic anatomy of a contractile unit of muscle.

- Explain the actomyosin complex theory.

- Detail a single muscle contraction.
- Analyze the major muscles according to their origin, insertion, and function.
- Distinguish between antagonistic and synergistic muscles.
- Diagram and label a neuromuscular synapse.
- Describe the motor end-plate.

III. IMPORTANT TERMS

Using your textbook, define the following terms:

acetylcholine (ah-seet-il-ko'-leen) _____

_____

actin (ack'-tin) _____

_____

antagonist (an-tag'-ah-nist) _____

_____

aponeuroses (ap-o-new-ro'-sis) _____

_____

ATP (ay-tee-pee) _____

_____

axon (acks'-on) _____

_____

contraction (kon-track'-shun) _____

_____

cross-bridges (kros bridj-es) _____

_____

elasticity (ee-las-tis'-it-ee) _____

_____

energy (en'-ur-jee) _____

_____

fascia (fash'-uh) _____

_____

fatigue (fah-teeg') _____

_____

fibrillation (fi-bril-lay′-shun) _____

_____

hypertrophy (hi-pur′-trah-fee) _____

_____

insertion (in-sur′-shun) _____

_____

muscle (mus′-ul) _____

_____

myalgia (mi-al′-juh) _____

_____

myofibril (mi-o-fi′-bril) _____

_____

myosin (mi′-o-sin) _____

_____

origin (or′-ah-jin) _____

_____

peristalsis (per-ah-stal′-sis) _____

_____

sarcomere (sahr′-ko-meer) _____

_____

spasm (spaz′-um) _____

_____

synergist (sin′-ur-jist) _____

_____

tendon (ten′-dun) _____

_____

vesicles (ves′-i-kuls) _____

_____

IV. EXERCISES

Complete the following exercises in the order given. A precise set of terms
and diagrams has been chosen to describe the muscular system.

Exercise 7.1

Labeling.   Write the name of the structure of a muscle fiber in the space
provided. Color the bands differently.

Key:

A-band(s)

cross-bridges

energy

H-zone

I-band(s)

sarcomere

Z-line

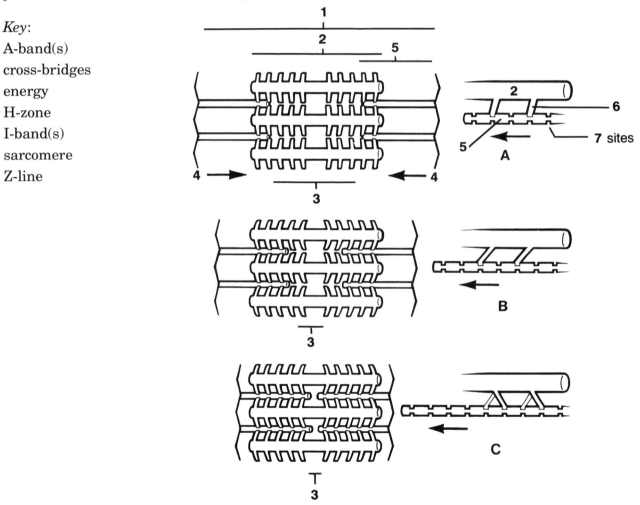

Figure 7.1   Contraction of a muscle. **A.** Sliding filament theory proposes that the
—2— contain flexible —6— that come in contact with —7— sites on the more
numerous —5— . **B.** and **C.** With the availability of —7— , the —6— pull the
active filament a short distance (**B**), release it, and attach to another site (**C**),
resulting in a shortening of the —3— between the —5—: contraction.

1. _____

2. _____

3. _____

4. _____

5. _____

6. _____

7. _____

Exercise 7.2

Labeling.   Write the name of the structure in the space provided. Color the nerve fiber one color and the muscle fiber another.

Key:

acetylcholine

axon

muscle

myelin

Schwann

terminal

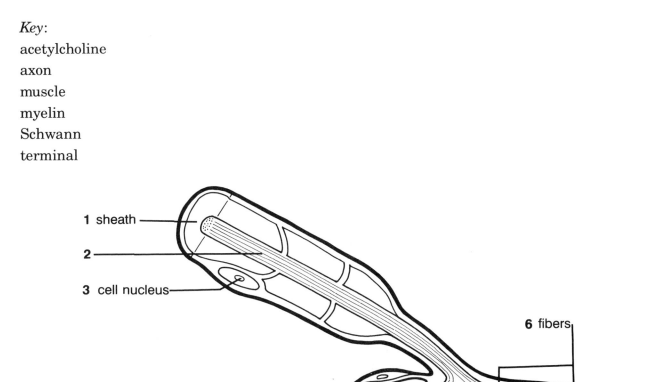

1 sheath

2

3 cell nucleus

4 vesicles

5 fiber

6 fibers

Figure 7.2   Cross-section of a motor end-plate on a skeletal muscle, showing the actual connection between the muscle fiber and the terminal branch of the nerve fiber.

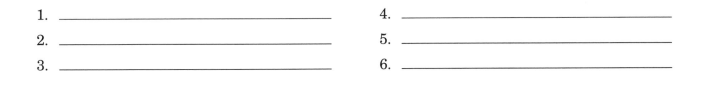

1. _____

2. _____

3. _____

4. _____

5. _____

6. _____

Exercise 7.3

Labeling. Write the name of the muscle on the space provided. Color the muscle mass leaving the tendons white.

Key:

adductor
biceps
deltoid
extensor
external
flexor
frontalis
gastrocnemius
gracilis
oculi
orbicularis
pectoralis
rectus
sartorius
serratus
sternocleidomastoid
tibialis
trapezius
vastus
zygomaticus

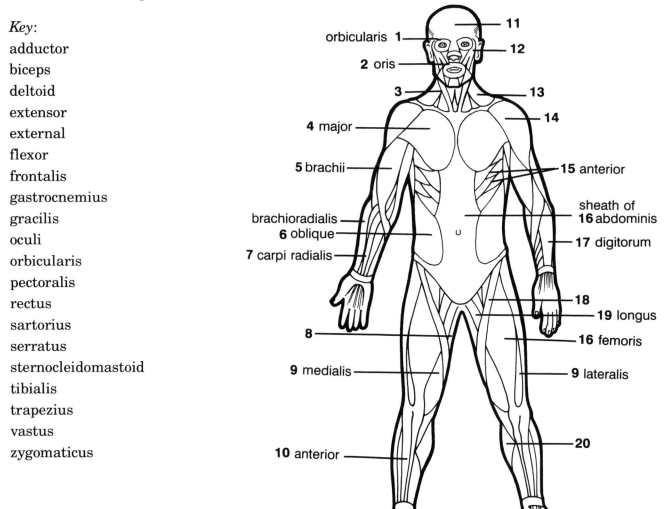

Figure 7.3   Muscles of the body. Anterior aspect.

1. _____
2. _____
3. _____
4. _____
5. _____
6. _____
7. _____
8. _____
9. _____
10. _____

11. _____
12. _____
13. _____
14. _____
15. _____
16. _____
17. _____
18. _____
19. _____
20. _____

Exercise 7.4

Labeling. Write the name of the muscle in the space provided. Color the muscle mass leaving the tendons white.

Key:

Achilles

biceps

deltoid

gastrocnemius

gluteus

infraspinatus

latissimus

occipitalis

plantaris

semimembranosus

semitendinosus

soleus

teres

trapezius

triceps

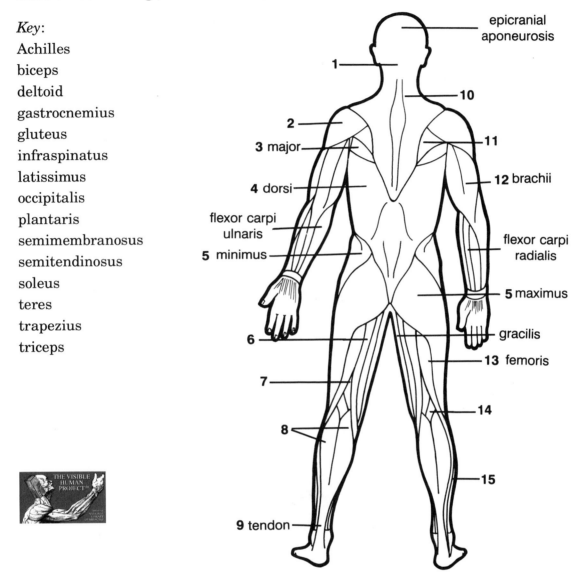

Figure 7.4 Muscles of the body. Posterior aspect.

1. _____	9. _____
2. _____	10. _____
3. _____	11. _____
4. _____	12. _____
5. _____	13. _____
6. _____	14. _____
7. _____	15. _____
8. _____	

V. TEST ITEMS

A. *Multiple Choice.*   There is only one answer that is either correct or most appropriate. Circle the answer that corresponds to the question.

1. Which of the following are common to all three types of muscle—skeletal, cardiac, and smooth?
   a. Contraction must be initiated by an action potential in a nerve fiber to the muscle cell.
   b. The interaction of actin, myosin, and ATP is responsible for generating force by the muscle cells.
   c. Actin potentials are conducted from cell to cell through cell junctions known as synapses.

2. The connective tissue component of skeletal muscle that surrounds fasciculi is called the
   a. perimysium.   c. endomysium.
   b. epimysium.   d. tendomysium.

3. The ability of muscle tissue to receive and respond to a stimulus is referred to as
   a. contractility.   c. elasticity.
   b. irritability.   d. extendibility.

4. Which of the following groupings is incorrect?
   a. skeletal, striated, voluntary
   b. smooth, visceral, involuntary
   c. cardiac, striated, voluntary
   d. cardiac, striated, involuntary

5. A sarcolemma is present in
   a. epithelial cells.   c. adipose tissue.
   b. striated muscle tissue.   d. blood.

6. According to the sliding-filament model of muscle action
   a. cross-bridges are lateral extensions of thick filaments, but temporarily attach to thin ones.
   b. cross-bridges are on thin filaments, but temporarily attach to thick ones.
   c. cross-bridges are on both thick and thin filaments.
   d. only filaments of the same size are interconnected.

7. When muscle fibers contract, the H-zone
   a. increases as the actin moves.
   b. decreases as the filaments slide.
   c. increases as the myosin disappears.
   d. decreases as the myosin enlarges.

8. Whole skeletal muscles
   a. are found in and about internal organs.
   b. work in antagonistic pairs.
   c. get longer when they contract.
   d. contain calcium deposits.
   e. all of these

9. Microscopically, muscle fibers contain parallel myofibrils, banded by repeating units. Each unit is called
   a. an actin.
   b. a myosin.
   c. a sarcomere.
   d. a myofibril.

10. Usually every normal muscle has at least some
    a. tone.
    b. tetanus.
    c. flexors.
    d. extensors.

11. Which of these gives the correct order from large to small?
    a. muscle, muscle cells, myofibrils, sarcomeres, myosin filaments, actin filaments
    b. muscle, muscle fibers, sarcomeres, myosin filaments, actin filaments, myofibrils
    c. muscle, sarcolemma, myofibrils, myosin filaments, actin filaments
    d. muscle cells, myofibrils, sarcoplasm, filaments

12. Intense muscular activity usually results in
    a. oxygen debt.
    b. muscle fatigue.
    c. accumulation of lactic acid.
    d. increased respiration.
    e. all of these

13. Each muscle fiber in striated muscle contains
    a. cilia.
    b. myofibrils.
    c. muscle hairs.
    d. villi.

14. During muscle fatigue, the production of lactic acid
    a. indicates lack of O_2.
    b. accompanies alcohol formation.
    c. is responsible for carbon dioxide formation.
    d. is necessary for glucose formation.
    e. occurs in the citric acid cycle.

15. The decrease in the force of muscular contraction after a period of repeated stimuli is called
    a. treppe.
    b. fatigue.
    c. contracture.
    d. none of the above

16. Complete fusion of muscle twitches to give a sustained contraction is called
    a. tetanus.
    b. summation.
    c. treppe.
    d. contracture.

17. At the myoneural junction
    a. acetylcholine is released from the muscle cell in response to an action potential.
    b. curare prevents the release of acetylcholine in response to an action potential.
    c. acetylcholine is rapidly broken down by an enzyme present in the end-plate membrane.
    d. the end-plate potential results in the inside of the membrane at the end-plate becoming positive and the outside positive.

18. According to the sliding-filament hypothesis
    a. potassium ions are necessary for contraction.
    b. Z-lines move away from the A-band.
    c. actin filaments move toward each other.
    d. myosin filaments move toward each other.

19. The following clinical symptoms—degeneration of muscle fibers, muscle atrophy, weakening of skeletal muscles, and fat deposition—are associated with
    a. muscular dystrophy.   c. convulsions.
    b. fatigue.   d. myositis.

20. Myofibrils are stacked in definite compartments partitioned by separations called Z-lines. Such compartments are known as
    a. sarcoplasm.   c. triads.
    b. sarcoplasmic reticulum.   d. sarcomeres.

B. *Matching Questions.*   Each of the phrases in COLUMN B refers to a word or phrase in COLUMN A. Insert the letter of the word or phrase that best describes it. Some words or phrases may be used more than once or not at all.

Column A	Column B
1. ___ myofilaments make up a	a. actin
2. ___ contain the protein myosin	b. troponin
3. ___ attached to the Z-line	c. myofibril
4. ___ a regulator protein	d. smooth muscle
5. ___ a single muscle cell	e. recruitment
6. ___ the combination of the motor neuron and a muscle fiber	f. cardiac muscle
7. ___ the process of increasing the number of motor neurons	g. norepinephrine
8. ___ lacks a well-developed sarcoplasmic reticulum	h. muscle fiber
9. ___ has a well-developed sarcoplasmic reticulum	i. a band
10. ___ a neurotransmitter	j. motor unit

Column A	Column B
1. ___ deltoid	a. extends thigh and flexes leg
2. ___ hamstring group	b. flexes leg and extends foot
3. ___ triceps	c. extends leg and flexes thigh
4. ___ gastrocnemius	d. flexes and adducts arm
5. ___ quadriceps femoris group	e. extends forearm
6. ___ pectoralis major	f. abducts arm

C. *True-False.*   Place a *T* or *F* in the space provided.

____ **1.** The thin filaments of a myofibril are composed of the protein myosin.

____ **2.** The motor unit is the basic unit of contraction.

____ **3.** The A-band contains myosin and is light.

____ **4.** The Z-line is located in the central region of the I-band.

____ **5.** Muscle tone results from a constant stream of stimuli to the skeletal muscles.

____ **6.** Acetylcholine also plays an important role in ATP synthesis.

____ **7.** A muscle fiber is made up of myofibrils, which in turn are made up of overlapping thick and thin protein filaments.

____ **8.** When skeletal muscle contracts, the A-band decreases in length.

____ **9.** Injuring your biceps brachii muscle will impair your ability to flex your forearm.

____ **10.** The point of attachment of a muscle to the bone that is moved as the muscle contracts is called the insertion of the muscle.

____ **11.** With the vertebrae, ribs, and pelvis fixed, contraction of the latissimus dorsi will extend and abduct the arm.

____ **12.** Smooth muscle is found primarily in the musculature of the extremities.

____ **13.** Smooth muscles are arranged in sheets and in less organized patterns.

____ **14.** A disorder in which the muscle shortens in length is hypertrophy.

____ **15.** Myalgia refers to pain in the muscular tissue.

____ **16.** Synergists are muscles that assist the agonist.

____ **17.** According to the all-or-none principle, muscle fibers contract all the way or they do not contract at all.

____ **18.** On a myogram of a twitch contraction, the relaxation period is indicated as an upward tracing.

____ **19.** Fibrositis of the lower back is called myopathy.

____ **20.** The biceps is a flexor and the triceps is an extensor; therefore, they are antagonists.

Answer Sheet—Chapter 7

Exercise 7.1

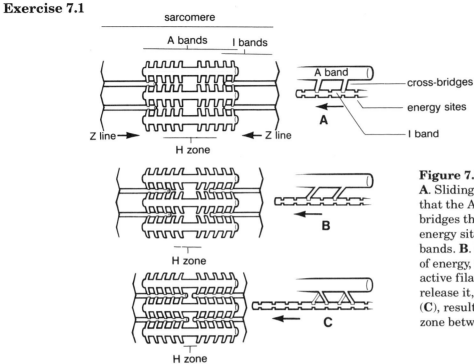

Figure 7.1 Contraction of a muscle. **A**. Sliding filament theory proposes that the A-bands contain flexible cross-bridges that come in contact with energy sites on the more numerous I-bands. **B**. and **C**. With the availability of energy, the cross-bridges pull the active filament a short distance (**B**), release it, and attach to another site (**C**), resulting in a shortening of the H-zone between the I-bands: contraction.

Exercise 7.2

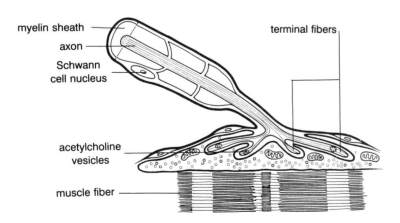

Figure 7.2 Cross-section of a motor end-plate on a skeletal muscle, showing the actual connection between the muscle fiber and the terminal branch of the nerve fiber.

Exercise 7.3

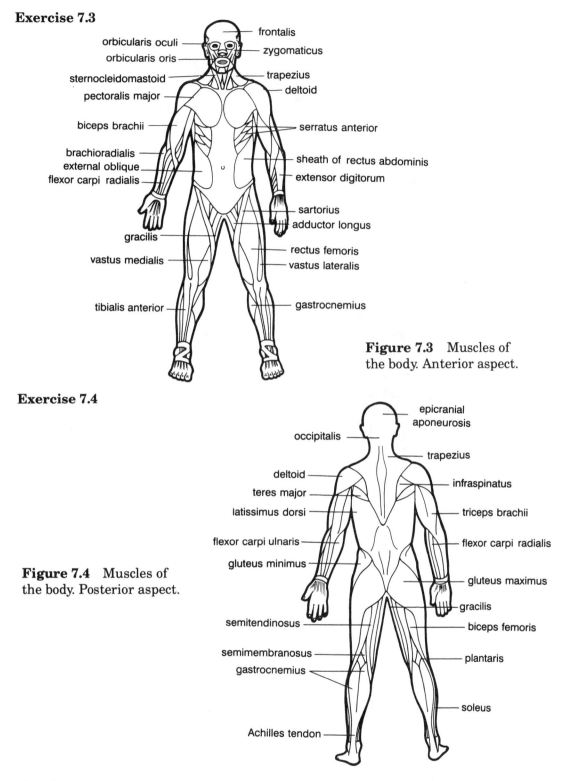

Figure 7.3 Muscles of the body. Anterior aspect.

Exercise 7.4

Figure 7.4 Muscles of the body. Posterior aspect.

Test Items

A. 1.b, 2.a, 3.a, 4.c, 5.b, 6.a, 7.b, 8.b, 9.c, 10.a, 11.a, 12.e, 13.b, 14.a, 15.b, 16.a, 17.c, 18.c, 19.a, 20.d.

B. 1.c, 2.i, 3.a, 4.b, 5.h, 6.j, 7.e, 8.d, 9.f, 10.g.
   1.f, 2.a, 3.e, 4.b, 5.c, 6.d.

C. 1.F, 2.T, 3.F, 4.T, 5.T, 6.F, 7.T, 8.F, 9.T, 10.T, 11.F, 12.F, 13.T, 14.F, 15.T, 16.T, 17.T, 18.F, 19.F, 20.T.

The Muscles

Across

3   neurotransmitter substance

7   a sheet of connective tissue

8   the ability to work

11   inability to respond to a stimulus

13   the contractile unit of a myofibril

15   muscle pain

16   connective tissue that connects muscle to bone

17   muscle protein found on the A-band

Down

1   increase in size of an organ or tissue

2   an opposite action

3   muscle protein found on the I-band

4   ability to stretch and return to normal

5   a muscle becomes short and thick

6   attachment of a muscle to the less movable bone

9   an involuntary muscle contraction

10   two or more structures working together

11   uncoordinated muscle contraction

12   attachment of a muscle to the more movable bone

14   contractile fibers within a muscle fiber

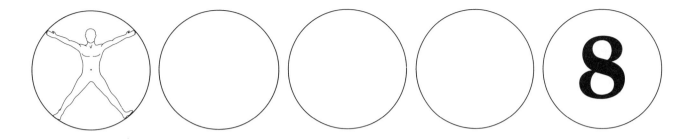

The Voluntary
Nervous System

I. CHAPTER SYNOPSIS

The nervous system correlates, coordinates, and reacts to impulses transmitted by sensory receptors known as nerves. The structural features of nervous tissue are described for nerve cells and supporting neurological cells. Neurons are classified functionally as afferent, efferent, or central in relation to the CNS.

The nervous system is divided into two parts: the central nervous system (CNS) and the peripheral nervous system (PNS). The CNS consists of the brain inside the skull and the spinal cord, which runs up inside the vertebral column and expands into the brain. Communicating centers within the CNS and various nerve tracts make possible the appropriate unconscious or conscious response to sensory stimulus. The PNS is made up of a network of nerves and sense organs that gathers information from the rest of the body and feeds it into the brain.

Once some of the body systems have been studied, it becomes fairly obvious that these systems cannot function alone. The systems are interdependent. All must work together as one functioning unit to maintain homeostasis. The mechanism that ensures that the organs and systems operate in smooth coordination is the nervous system. Conditions within and outside the body are constantly changing, and one purpose of the nervous system is to respond to these internal and external changes so that the body may adapt itself.

II. OBJECTIVES

After reading the chapter, the student should be able to:

- Describe the structure of a neuron.
- Classify neurons according to their structure, function, and position.
- Explain a simple reflex.
- Distinguish between a simple versus complex reflex.
- Differentiate between the white and gray matter of nerve tissue.
- Describe the cerebrum, and identify the lobes by their location and function.
- List the organs of the hind brain and relate their functions to the brain and spinal cord.
- Identify the twelve cranial nerves by their number, function, and distribution.

III. IMPORTANT TERMS

Using your textbook, define the following terms:

afferent (af'-ah-rent) _____

_____

arc (ark) _____

_____

association (ah-so-see-ay'-shun) _____

_____

axon (acks'-on) _____

_____

conduction (kahn-duk'-shun) _____

_____

cortex (kor'-teks) _____

_____

dendrite (den'-dright) _____

_____

dorsal (dor'-sul) _____

_____

efferent (ef'-ah-rent) _____

_____

epineurium (ep-ah-nyoor'-ee-um) _____

_____

fissure (fish'-ur) _____

_____

gray (gray) _____

_____

gyrus (ji'-rus) _____

_____

internuncial (int-ur-nun'-see-ul) _____

_____

medulla (mah-dul'-ah) _____

_____

motor (mote'-ur) _____

_____

myelin (mi'-ah-lin) _____

_____

neuralgia (nyoo-ral'-jah) _____

_____

neuritis (nyoo-right'-us) _____

_____

neurofibril (nyoor-o-fibe'-ril) _____

_____

neurolemma (nyoor-o-lem'-ah) _____

_____

plexus (plek'-sus) _____

_____

receptor (re-sep'-tur) _____

_____

reflex (ree'-fleks) _____

_____

sensory (sens'-ah-ree) _____

_____

spinal (spine'-al) _____

_____

ventral (ven'-trahl) _____

_____

IV. EXERCISES

Complete the following exercises in the order given. A precise set of terms and diagrams has been chosen to describe the nervous system.

Exercise 8.1

Labeling. Write the name of the structure in the space provided. Color the three types of nerves differently.

Key:

axon	neurilemma
cell body	neurofibrils
dendrite	Nissl
epineurium	nucleus
fasciculus	perineurium
muscle	Ranvier
myelin	receptors
myoneural	Schwann
nerve	

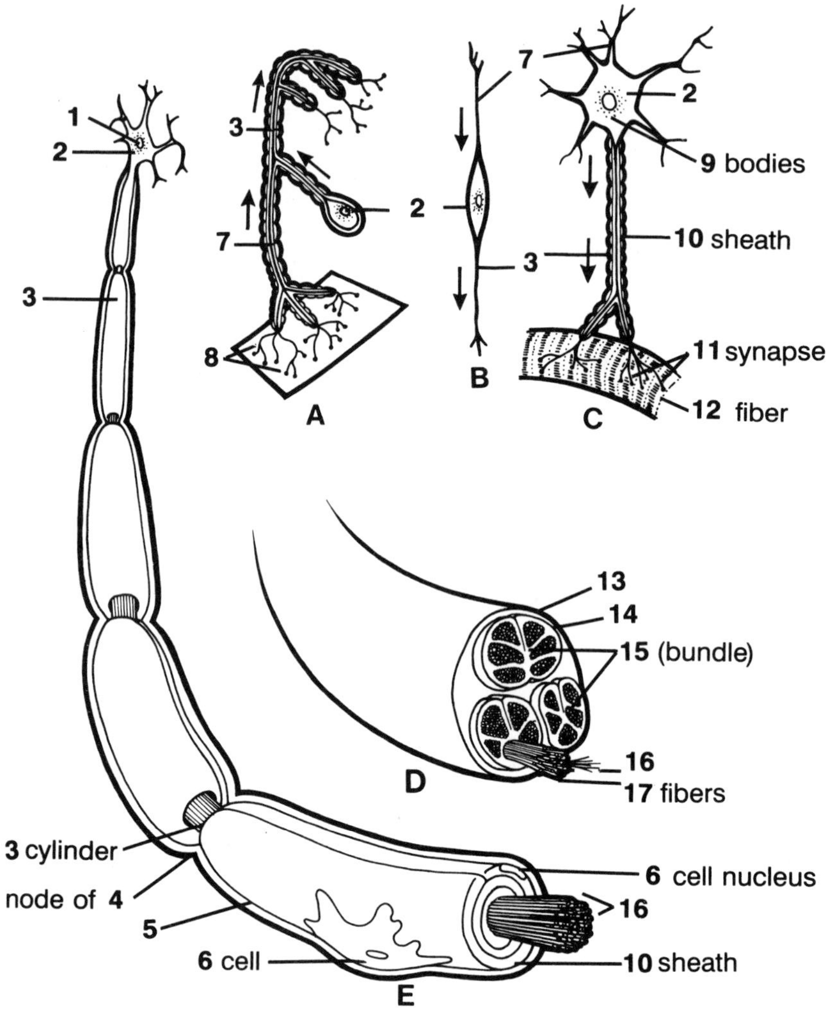

1. _____
2. _____
3. _____
4. _____
5. _____
6. _____
7. _____
8. _____
9. _____
10. _____
11. _____
12. _____
13. _____
14. _____
15. _____
16. _____
17. _____

Figure 8.1   The structure of a neuron. **A**. Monopolar sensory neuron. **B**. Bipolar retinal neuron in the eye. **C**. Multipolar motor neuron, showing a muscle-nerve synapse. **D**. Structure of a typical nerve. **E**. An enlarged extension of an axon, showing its protective myelin sheath and neurilemma.

Exercise 8.2

Labeling.   Write the name of the structure in the space provided. Color the white matter differently from the gray matter.

Key:

dorsal	internuncial	sensory
effector	motor	ventral
gray	receptor	white

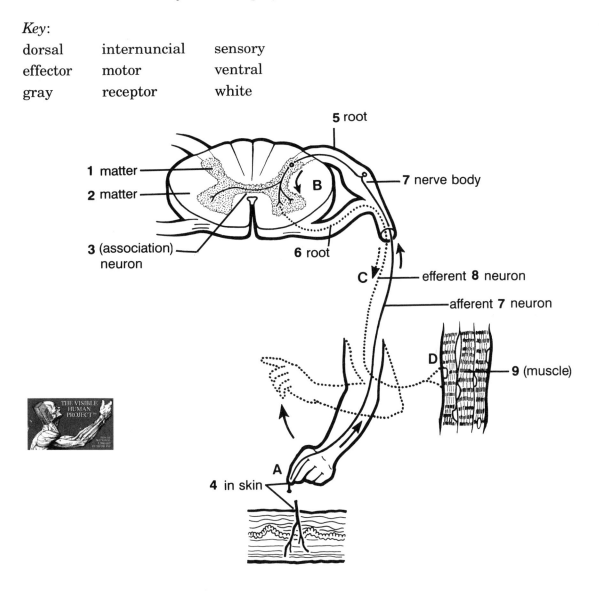

Figure 8.2   Diagram demonstrating a simple reflex arc. An impulse is initiated at the finger by a receptor (**A**) in the skin. The impulse travels over the sensory afferent neuron to the spinal cord, where it is transmitted via the internuncial neuron (**B**) to the efferent motor neuron. The impulse then travels via the efferent motor neuron (**C**) to the muscle (**D**), which effects a response.

1. _____

2. _____

3. _____

4. _____

5. _____

6. _____

7. _____

8. _____

9. _____

Exercise 8.3

Labeling. Write the name of the structure or function in the space provided. Color each lobe differently. Color lobes in A and B the same way.

Key:
association
auditory
central sulcus
cerebellum
cerebrum
frontal
lateral sulcus
medulla
motor
motor speech
occipital
parietal
pons
sensory
speech
spinal cord
taste
visual

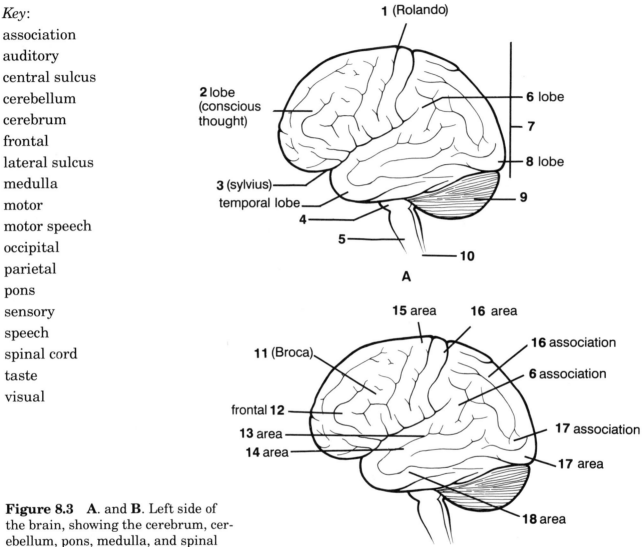

1 (Rolando)

2 lobe (conscious thought)

6 lobe

7

8 lobe

3 (sylvius)
temporal lobe
4

5

9

10

A

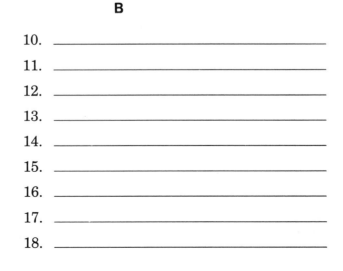

15 area   16 area

11 (Broca)

16 association

6 association

frontal 12

13 area

14 area

17 association

17 area

18 area

B

Figure 8.3 **A.** and **B.** Left side of the brain, showing the cerebrum, cerebellum, pons, medulla, and spinal cord.

1. _____
2. _____
3. _____
4. _____
5. _____
6. _____
7. _____
8. _____
9. _____

10. _____
11. _____
12. _____
13. _____
14. _____
15. _____
16. _____
17. _____
18. _____

Exercise 8.4

Labeling. Write the name of the cranial nerve in the space provided. Color the nerves according to their function.

Key:

abducens	oculomotor
accessory	olfactory
acoustic	optic
facial	trigeminal
glossopharyngeal	trochlear
hypoglossal	vagus

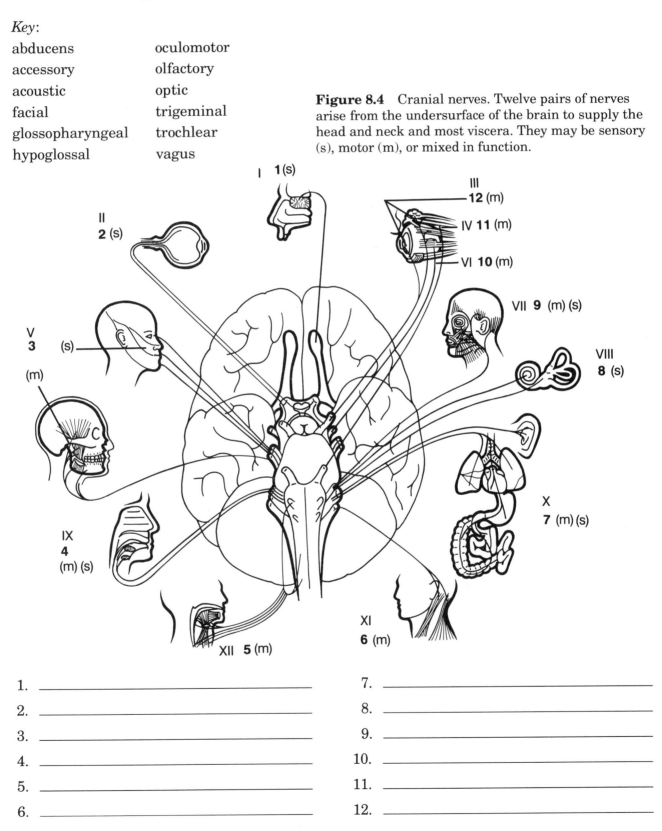

Figure 8.4 Cranial nerves. Twelve pairs of nerves arise from the undersurface of the brain to supply the head and neck and most viscera. They may be sensory (s), motor (m), or mixed in function.

I **1** (s)

III **12** (m)

IV **11** (m)

VI **10** (m)

II **2** (s)

VII **9** (m) (s)

VIII **8** (s)

V **3** (s) (m)

X **7** (m) (s)

IX **4** (m) (s)

XI **6** (m)

XII **5** (m)

1. _____
2. _____
3. _____
4. _____
5. _____
6. _____
7. _____
8. _____
9. _____
10. _____
11. _____
12. _____

V. TEST ITEMS

A. *Multiple Choice.*   There is only one answer that is either correct or most appropriate. Circle the answer that corresponds to the question.

1. If your body had to react immediately to stress, which system would assume control?
   a. nervous
   b. circulatory
   c. endocrine
   d. integumentary

2. Which of the following is not part of a neuron?
   a. cell body
   b. dendrite
   c. axon
   d. neuroeffector junction

3. A neuron that has one long axon and multiple short highly branched dendrites extending from the cell body is known as
   a. multipolar cell.
   b. bipolar cell.
   c. unipolar cell.
   d. ganglion.

4. A reflex arc
   a. always includes a sensory neuron and a motor neuron.
   b. always has its center in the brain or spinal cord.
   c. always terminates in muscle or gland.
   d. all of the preceding

5. The knee jerk in response to a sharp tap over the patellar tendon
   a. is an autonomic reflex.
   b. is a conditional reflex.
   c. is mediated by a three-neuron reflex arc.
   d. has its reflex center in the spinal cord.
   e. more than one of the above

6. If the anterior root of a spinal nerve were cut, what would be the result in the regions supplied by that spinal nerve?
   a. complete loss of sensation
   b. complete loss of movement
   c. complete loss of sensation and movement
   d. complete loss of sensation, movement, and autonomic control of blood vessels and sweat glands

7. Stimulation of a receptor initiates what event?
   a. action potential
   b. glycogenesis
   c. nerve regeneration
   d. none of the above

8. A neuron that transmits a nerve impulse to the central nervous system is called a(n)
   a. motor neuron.
   b. sensory neuron.
   c. bipolar neuron.
   d. association neuron.

9. A physician informs you that a patient has a disorder of the central nervous system. Which part of the nervous system is involved?
   a. nerves in the forearm
   b. nerves to the heart
   c. brain and spinal cord
   d. sympathetic neurons

10. Neurons that conduct impulses to the spinal cord or brain stem are called
    a. afferent neurons.
    b. efferent neurons.
    c. interneurons.
    d. visceral neurons.

11. The distal ends of sensory neuron dendrites are called
    a. effectors.
    b. Nissl bodies.
    c. receptors.
    d. synapses.

12. Which of the following is (are) present in neurons?
    a. Golgi apparatus
    b. mitochondria
    c. Nissl bodies
    d. all of the above

13. The part of a neuron that conducts impulses away from its cell body is called
    a. an axon.
    b. a dendrite.
    c. an effector.
    d. a neurilemma.

14. Which one of the following fissures separates the temporal lobe from the frontal and parietal lobes of the centrum?
    a. longitudinal
    b. lateral (fissure of Sylvius)
    c. central
    d. parietooccipital

15. Beneath the cerebral cortex is white matter composed of neuron fibers. Fibers that connect different areas of the cortex within a single hemisphere are called
    a. association fibers.
    b. commissures.
    c. projection fibers.
    d. all of the above

16. To expose the corpus callosum of the cerebrum, which landmark would you use to make the incision?
    a. longitudinal fissure
    b. central sulcus
    c. lateral cerebral sulcus
    d. parietooccipital sulcus

17. Abnormal body movements, such as uncontrollable shaking, and involuntary movements of the skeletal muscles, probably indicate damage to the
    a. sensory areas of the cerebrum.
    b. basal ganglia (cerebral nuclei).
    c. association areas of the cortex.
    d. primary olfactory areas.

18. A patient with a tumor of the cerebellum would probably exhibit
    a. absence of the patellar reflex.
    b. unconsciousness.
    c. the inability to execute smooth, precise movements.
    d. the inability to perform voluntary movements.

19. The vital centers for the control of heartbeat, respiration, and the control of blood vessel diameter are located in the
    a. pons.   c. cerebrum.
    b. medulla.   d. cerebellum.

20. The reason that the motor areas of the right cerebral cortex control voluntary movements on the left side of the body is that the
    a. cerebrum contains projection fibers.
    b. cerebellum controls voluntary movements.
    c. medulla contains decussating pyramids.
    d. pons connects the spinal cord with the brain.

B. *Matching Questions.* Each of the phrases in COLUMN B refers to a word or phrase in COLUMN A. Insert the letter of the word or phrase from COLUMN B that best describes it. Some words or phrases may be used more than once or not at all.

Column A		*Column B*
1. ___ coma		a. acute inflammation of the brain caused by a virus
2. ___ analgesia		b. abnormality of one or more vertebral arches in which part of the spinal cord may be exposed
3. ___ sciatica		
4. ___ paralysis		c. inflammation of a nerve
5. ___ torpor		d. attacks of pain along the entire course or branch of a peripheral sensory nerve
6. ___ shingles		
7. ___ neuritis		e. abnormally deep unconsciousness with an absence of voluntary responses to stimuli
8. ___ anesthesia		f. diminished or complete loss of ability to comprehend and/or express spoken or written words
9. ___ bacterial meningitis		
10. ___ viral encephalitis		g. abnormal inactivity or no response to normal stimuli
11. ___ spina bifida		h. severe pain along the sciatic nerve and its branches
12. ___ neuralgia		i. loss of feeling
13. ___ aphasia		j. insensibility to pain
		k. acute inflammation of the meninges caused by a bacterium
		l. diminished or total loss of motor function resulting from damage to nervous or muscular tissue
		m. inflammation caused by a virus that attacks sensory cell bodies of dorsal root ganglia and produces a characteristic line of skin blisters

Match the following malfunctions with the most probable cranial nerves. For each defect, you should pick the *single* most likely nerve.

Body Malfunctions

Cranial Nerves

1. ___ loss of hearing and the sense of equilibrium

2. ___ inability to roll the eyeball upward

3. ___ blindness

4. ___ inability to move the tongue

5. ___ inability to move the shoulder and turn the head

6. ___ loss of smell

7. ___ inability to focus and partial loss of eye movement

8. ___ inability to move the eyeball laterally

9. ___ difficulty in swallowing

10. ___ inability to masticate and lack of senses of the face

11. ___ inability to smile

12. ___ gastrointestinal problems

13. ___ no sense of taste from the posterior portion of the tongue

a. olfactory

b. optic

c. oculomotor

d. trochlear

e. trigeminal

f. abducens

g. facial

h. vestibulocochlear

i. glossopharyngeal

j. vagus

k. accessory

l. hypoglossal

C. *True-False.*   Place a *T* or *F* in the space provided.

___ 1. The nervous and endocrine systems play the major role in regulation and coordination of the body.

___ 2. The efferent pathway in a physiological control system is the path from the integrating center to an effector.

___ 3. The posterior columns of gray matter contain ventral horn cells.

___ 4. The lateral gray columns contain cell bodies of the axons that pass out to sympathetic ganglia.

___ 5. The anterior division of the thoracic spinal nerves form the thoracic plexus.

___ 6. The spinal cord of the adult extends the entire length of the vertebral canal.

___ 7. The posterior and anterior columns of the central gray matter together with the core follows the general form of the letter H.

___ 8. A reflex is a voluntary response to a harmful stimulus.

_____ **9.** Transection of the spinal cord results in a release of functions normally inhibited by the higher centers.

_____ **10.** The myelin sheath forms the continuous covering of a neuron.

_____ **11.** Some neurons are specialized for the secretion of hormonal substances.

_____ **12.** The neurons of the peripheral ganglia are true bipolar cells.

_____ **13.** Both bipolar and multipolar neurons have single axons.

_____ **14.** All cells have membrane potentials.

_____ **15.** All action potentials within any one cell obey the all-or-none law.

_____ **16.** The irritable and conducting units of the nervous system are called neurons.

_____ **17.** The Nissl bodies in the cytoplasm of neurons are active sites of protein synthesis.

_____ **18.** Motor nerve cells and interneurons are multipolar.

_____ **19.** Neurological cells are conducting units.

_____ **20.** Nerve endings for pain and for posture have a short period of adaptation.

Answer Sheet—Chapter 8

Exercise 8.1

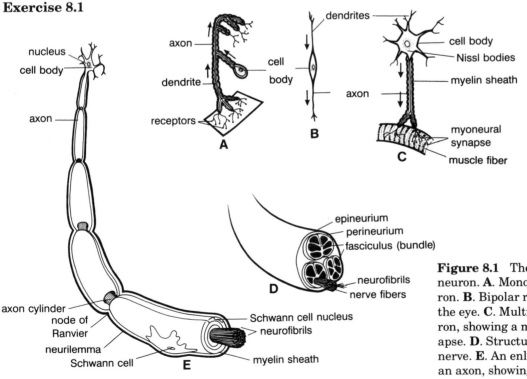

Figure 8.1 The structure of a neuron. **A.** Monopolar sensory neuron. **B.** Bipolar retinal neuron in the eye. **C.** Multipolar motor neuron, showing a muscle-nerve synapse. **D.** Structure of a typical nerve. **E.** An enlarged extension of an axon, showing its protective myelin sheath and neurilemma.

Exercise 8.2

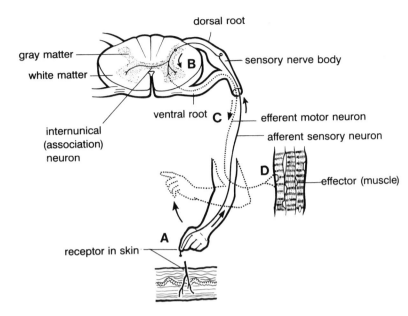

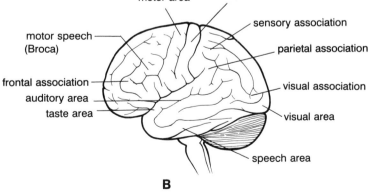

Figure 8.2 Diagram demonstrating a simple reflex arc. An impulse is initiated at the finger by a receptor (**A**) in the skin. The impulse travels over the sensory afferent neuron to the spinal cord, where it is transmitted via the internuncial neuron (**B**) to the efferent motor neuron. The impulse then travels via the efferent motor neuron (**C**) to the muscle (**D**), which effects a response.

Exercise 8.3

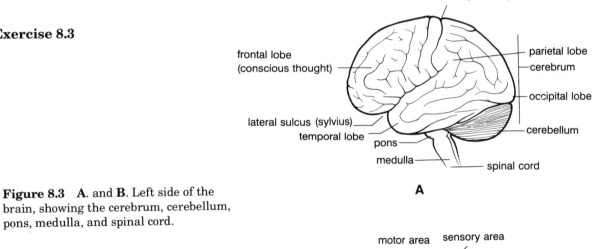

Figure 8.3 A. and **B**. Left side of the brain, showing the cerebrum, cerebellum, pons, medulla, and spinal cord.

Exercise 8.4

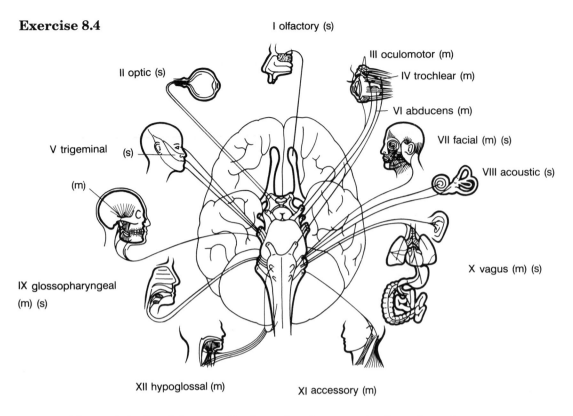

Figure 8.4 Cranial nerves. Twelve pairs of nerves arise from the undersurface of the brain to supply the head and neck and most viscera. They may be sensory (s), motor (m), or mixed in function.

Test Items

A. 1.a, 2.d, 3.a, 4.d, 5.d, 6.b, 7.a, 8.b, 9.c, 10.a, 11.c, 12.d, 13.a, 14.b, 15.a, 16.a, 17.b, 18.c, 19.b, 20.c.

B. 1.e, 2.j, 3.h, 4.l, 5.g, 6.m, 7.c, 8.i, 9.k, 10.a, 11.b, 12.d, 13.f.
   1.h, 2.d, 3.b, 4.l, 5.k, 6.a, 7.c, 8.f, 9.i, 10.e, 11.g, 12.j, 13.l.

C. 1.T, 2.T, 3.F, 4.T, 5.F, 6.F, 7.T, 8.F, 9.T, 10.F, 11.T, 12.F, 13.T, 14.T, 15.T, 16.T, 17.T, 18.T, 19.F, 20.F.

Central Nervous System

Across

1 a connecting nerve

8 the base of the brain or top of the spinal cord

10 sensory nerve ending that initiates a stimulus

11 an area of non-myelinated nerves

14 deep cleft or groove

15 neuron that transmits an impulse to or toward the brain or spinal cord

18 an involuntary response to a stimulus

19 inflammation of a nerve

21 pertaining to the front; anterior

22 flat, raised surface of the brain

Down

1 extension of a nerve that transmits an impulse to another nerve

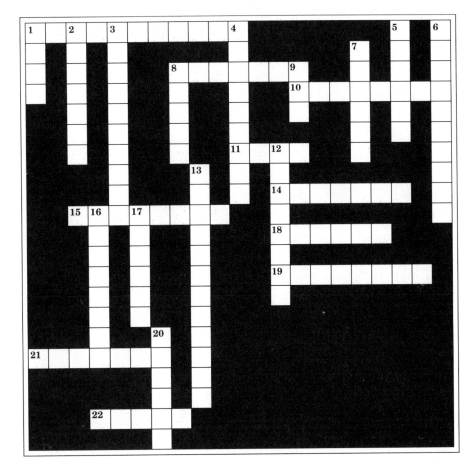

2 an area of interpretation; a feeling

3 transmission of an impulse from one area to another

4 pain along the nerve

5 the outer part of an organ

6 nerve cell membrane

7 connective tissue covering of a nerve

8 to act or move

9 beginning and ending at the same point

12 to or toward an organ or structure

13 a connecting neuron of the brain

16 away from an organ or area

17 pertaining to the back; posterior

20 network of nerves

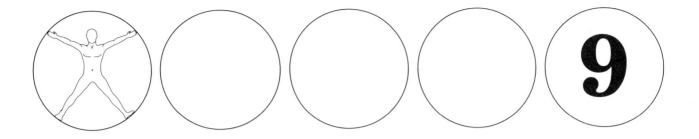

The Involuntary Nervous System

9

I. CHAPTER SYNOPSIS

Students are introduced to the involuntary nervous system (also known as the autonomic nervous system or ANS) and the structure and physiology of this system are described in detail. The autonomic system is the portion of the nervous system that regulates the activities of smooth muscle, cardiac muscle, and glands. Structurally, the system consists of visceral efferent neurons organized into nerves, ganglia, and plexuses. Functionally, it usually operates without conscious control. The autonomic nervous system consists of two principal divisions: the sympathetic and the parasympathetic. Both divisions are compared in terms of structure, physiology, and chemical transmitters released.

Physiologically, both divisions of the ANS contribute to the maintenance of homeostasis through their regulation of visceral activities. Most visceral effectors have a dual autonomic nerve supply from each division of the ANS. The parasympathetic division is concerned with activities of restoration and conservation of bodily energy and with the elimination of wastes. The sympathetic division is influenced by changes in external environment and is best known for the initiation of a widespread response to physical danger. Although visceral structures have a degree of functional independence, they are constantly monitored and regulated by cortical and subcortical areas, primarily the hypothalamus. Biofeedback is a term applied to techniques used to gain conscious control over visceral responses. Acetylcholine is released by cholinergic fibers that include all preganglionic fibers and the postganglionic fibers of the parasympathetic division. Norepinephrine is released by most postganglionic fibers of the sympathetic division, and fibers that release this neurotransmitter are called adrenergic fibers.

II. OBJECTIVES

After reading the chapter, the student should be able to:

- Describe the anatomy of the autonomic nervous system.

- Differentiate between a white ramus and gray ramus and their contents.

- Integrate the autonomic nervous system with the central nervous system.

- Describe the sympathetic trunk and its preganglionic and postganglionic fibers.

- Explain the chemical action of both the sympathetic and parasympathetic divisions.

- Identify adrenergic and cholinergic fibers and their receptors.

- Define antagonism.

III. IMPORTANT TERMS

Using your textbook, define the following terms:

adrenergic (ad-rah-nur´-jik) _____

_____

autonomic (awt-uh-nom´-ik) _____

_____

celiac (see´-lee-ak) _____

_____

cholinergic (ko-lah-nur´-jik) _____

_____

cholinesterase (ko-lah-nes´-tah-race) _____

_____

collateral (kah-lat´-ah-rul) _____

_____

communicans (kah-myoo´-neh-kanz) _____

_____

enteric (en-ter´-ik) _____

_____

epinephrine (ep-ah-nef´-rin) _____

_____

ganglion (gang´-glee-ahn) _____

_____

gray ramus (gray ray´-mus) _____

_____

hypogastric (hi-po-gas´-trik) _____

_____

hypothalamus (hi-po-thal´-ah-mus) _____

_____

inhibit (in-hib´-it) _____

_____

involuntary (in-vol´-un-ter-ee) _____

_____

medulla (mah-dul´-ah) _____

_____

neurotransmitter (nyoor-o-tranz-mit´-ur) _____

_____

norepinephrine (nor-ep-ah-nef´-rin) _____

_____

parasympathetic (par-ah-sim-pah-thet´-ik) _____

_____

peripheral (pah-rif´-ah-rul) _____

_____

plexus (plek´-sus) _____

_____

postganglionic (post-gang-glee-on´-ik) _____

_____

preganglionic (pree-gang-glee-on´-ik) _____

_____

spinal (spine´-al) _____

_____

sympathetic (sim-pah-thet´-ik) _____

_____

synoptic (sah-nop´-tik) _____

_____

target (tahr´-gut) _____

_____

vasoconstriction (vay-zo-kahn-strik´-shun) _____

_____

vasodilation (vay-zo-di-lay´-shun) _____

_____

vasomotor (vay-zo-mote´-ur) _____

_____

white ramus (wite ray´-mus) _____

_____

IV. EXERCISES

Complete the following exercises in the order given. A precise set of terms and diagrams has been chosen to describe the involuntary nervous system.

Exercise 9.1

Labeling.   Write the name of the structure in the space provided. Color the spinal cord and visceral cord differently.

Key:

bladder	parasympathetic	preganglionic
ganglion	pelvic	spinal
hypogastric	postganglionic	sympathetic

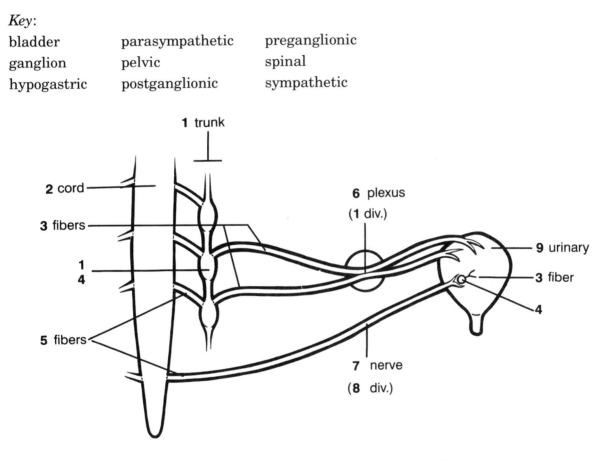

Figure 9.1   Diagram of the lumbosacral segment of the autonomic nervous system, showing the arrangement of ganglia, preganglionic fibers, and postganglionic fibers of both the sympathetic and para-sympathetic divisions to the spinal cord and visceral organs.

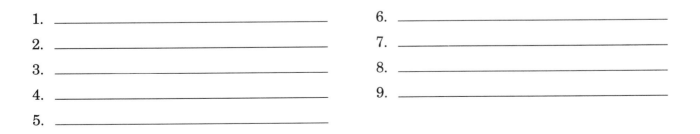

1. _____   6. _____

2. _____   7. _____

3. _____   8. _____

4. _____   9. _____

5. _____

Exercise 9.2

Labeling. Write the name of the structure in the space provided. Color the spinal cord, trunk, and visceral organ differently.

Key:

collateral	ganglion	root	trunk
communicans	gray	spinal	ventral
dorsal	postganglionic	sympathetic	white

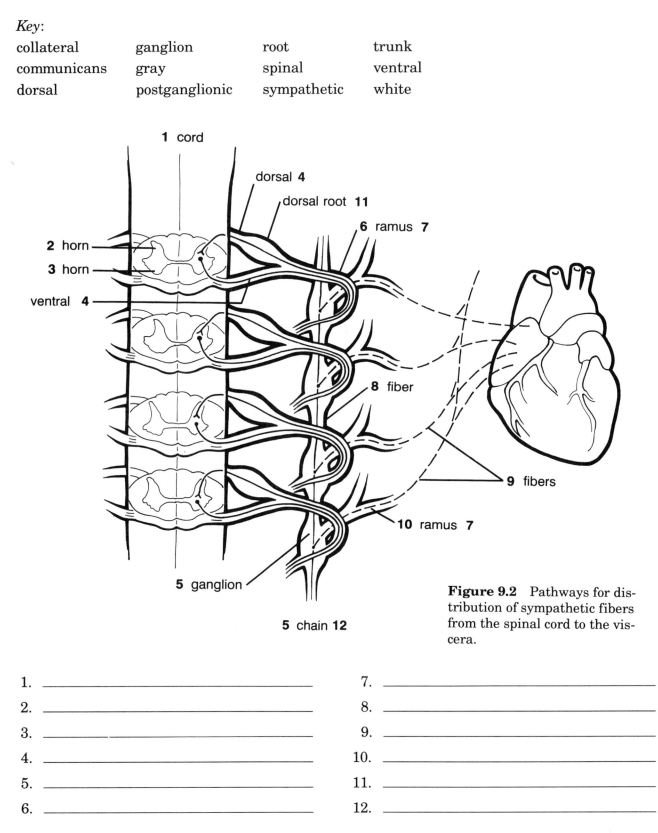

Figure 9.2 Pathways for distribution of sympathetic fibers from the spinal cord to the viscera.

1. _____
2. _____
3. _____
4. _____
5. _____
6. _____
7. _____
8. _____
9. _____
10. _____
11. _____
12. _____

Exercise 9.3

Completion. Fill in the blanks with the appropriate terms. Some terms may be used more than once.

Key:

adrenal
adrenergic
autonomic
bloodstream
chemical
epinephrine
excitation
gray
medulla
neurons
norepinephrine
peripheral
postganglionic
spinal
stimulate
sympathetic
terminal
viscera

Key:

acetic acid
acetylcholine
acts
bloodstream
both
chemistry
choline
cholinesterase
depolarizing
destroys
ganglia
hyperpolarization
inhibit
inhibition
not
parasympathetic
preganglionic
synaptic
target

CHEMICAL TRANSMITTERS

Adrenergic Fibers

The _____ filaments of most _____ post-ganglionic _____ produce chiefly _____. This substance is also secreted by the _____ of the _____ gland, and therefore these fibers are classi-fied as _____ . Exceptions are the sympathetic fibers to sweat glands, to the blood vessels of the skin, and to the arrector muscles that elevate the hair. These _____ fibers enter the _____ nerves through the _____ rami communicantes and reach the skin incor-porated in _____ nerves.

The effects of _____, also secreted by the adrenal medulla, and _____ can be widespread because these _____ substances, which result from _____ of the _____ postganglionic fibers, are carried by the _____ . _____ transmitters _____ the _____ controlled by the sympathetic division of the _____ nervous system.

Cholinergic Fibers

_____ fibers also produce a chemical substance, _____ , which is promptly converted to _____ and _____ by the action of the enzyme _____ . Acetylcholine is a _____ substance that aids _____ transmission at the _____ termi-nals in _____ sympathetic and parasympathetic _____ . At the postganglionic terminal of the para-sympathetic or craniosacral division, secretion of _____ to _____ the organ. The ability of acetylcholine to excite at the first synapse and inhibit at the second synapse is due to differences in the _____ of the _____ cells. In the heart, for example, parasym-pathetic stimulation is followed by _____ of the car-diac cells and _____ of the heartbeat. The enzyme _____ quickly _____ acetylcholine, which therefore has effects only locally, where it is secreted. Unlike norepinephrine, it probably is _____ carried by the _____ .

V. TEST ITEMS

A. *Multiple Choice.*   There is only one answer that is either correct or most appropriate. Circle the answer that corresponds to the question.

1. Which statement concerning the autonomic nervous system is *not* true?
   a. It usually operates without any conscious control.
   b. It regulates visceral activities.
   c. All of its axons are afferent fibers.
   d. It contains rami and ganglia.

2. Terminal ganglia receive
   a. postganglionic fibers from the parasympathetic division.
   b. postganglionic fibers from the sympathetic division.
   c. preganglionic fibers from the parasympathetic division.
   d. preganglionic fibers from the sympathetic division.

3. When the sympathetic nervous system is stimulated, which of the following occurs?
   a. Vessels in the skeletal muscles constrict.
   b. Blood pressure increases.
   c. Respirations decrease.
   d. Peristalsis increases.

4. When the parasympathetic nervous system is stimulated, which of the following occurs?
   a. Digestive processes are increased.
   b. Bronchioles dilate.
   c. Pupils dilate.
   d. Vessels in the skin constrict.
   e. more than one of the above

5. The chemical mediator that aids in the transmission of nerve impulses at the synapses is
   a. strychnine.   c. cholinesterase.
   b. monamine oxidase.   d. acetylcholine.

6. Autonomic nerve fibers supply
   a. skeletal muscle, cardiac muscle, and glands.
   b. visceral muscle, cardiac muscle, and glands.
   c. skeletal muscle, visceral muscle, and cardiac muscle.
   d. skeletal muscle, visceral muscle, and glands.

7. Sympathetic responses generally have widespread effects on the body because
   a. preganglionic fibers are short and postganglionic fibers are long.
   b. myoneural junctions contain a substance that inactivates acetylcholine.
   c. preganglionic fibers synapse with several postsynaptic fibers.
   d. they reach visceral effectors faster than parasympathetic impulses.

8. The cell bodies of preganglionic neurons of the parasympathetic division of the autonomic nervous system are located in the
   a. lateral gray horns of the thoracic cord.

   b. nuclei in the brain stem and lateral gray horns of the sacral cord.
   c. lateral gray horns of the cervical cord.
   d. lateral gray horns of the lumbar cord.

 9. Autonomic ganglia located on either side of the vertebral column from the base of the skull to the coccyx are called
   a. prevertebral ganglia.
   b. collateral ganglia.
   c. terminal ganglia.
   d. sympathetic trunk ganglia.

10. Axons from preganglionic neurons of the parasympathetic division of the autonomic nervous system
   a. synapse in sympathetic chain ganglia.
   b. synapse in prevertebral ganglia.
   c. synapse in terminal ganglia.
   d. are part of the thoracolumbar outflow.

11. In their course from vertebral rami to the sympathetic trunk, sympathetic preganglionic fibers are contained in structures called
   a. white rami communicantes.
   b. meningeal branches.
   c. dorsal rami.
   d. gray rami communicantes.

12. Diminished or total loss of motor function from damage to nervous tissue or a muscle is called
   a. paralysis.   c. neuralgia.
   b. sciatica.   d. aphasia.

13. Which of these is *not* a component of the sympathetic nervous system?
   a. white rami communicantes
   b. inferior cervical ganglion
   c. ciliary ganglion
   d. gray rami communicantes

14. Which statement is *not* true of the parasympathetic nervous system?
   a. It forms the craniosacral outflow.
   b. It contains terminal ganglia.
   c. Its ganglia are near or within visceral effectors.
   d. It is distributed throughout the body.

15. One sympathetic response is
   a. dilation of the bronchial tubes.
   b. normal or excessive lacrimal secretion.
   c. increased intestinal motility.
   d. increased pancreatic secretion.

16. Which of these is *not* true of the autonomic nervous system?
   a. It controls heartbeat, peristalsis, secretion of glands.
   b. It is composed of sympathetic and parasympathetic systems.
   c. It is composed only of fibers that inhibit the function of various organs of the body.
   d. The impulses require two motor neurons to reach their destination.

17. The autonomic nervous system
   a. is made up of motor neurons only.
   b. has preganglionic and postganglionic fibers.
   c. has cell bodies in both the cord and in ganglia.
   d. all of the above

18. Which of the following is not generally associated with the autonomic system?
   a. speaking
   b. digestion
   c. heartbeat
   d. body temperature

19. When a person's heart races
   a. the parasympathetic effect is more influential than the sympathetic effect.
   b. the sympathetic effect is more influential than the parasympathetic effect.
   c. there is a perfect balance between the parasympathetic and the sympathetic effects.
   d. voluntary control is in effect.

20. Parasympathetic fibers reach the intestine via the
   a. greater splanchnic nerve.
   b. phrenic nerve.
   c. hypoglossal nerve.
   d. vagus nerve.

B. *Matching Questions.* Each of the phrases in COLUMN B refers to a word or phrase in COLUMN A. Insert the letter of the word or phrase from COLUMN B that best describes it. Some words or phrases may be used more than once, or not at all.

	Column A	*Column B*
1. ___	preganglionic	**a.** thoracolumbar
2. ___	postganglionic	**b.** cholenergic
3. ___	epinephrine	**c.** craniosacral
4. ___	autonomic	**d.** heart
5. ___	enteric plexus	**e.** involuntary
6. ___	sympathetic trunk	**f.** voluntary
7. ___	parasympathetic	**g.** gray ramus
8. ___	acetylcholine	**h.** digestive tract
9. ___	solar plexus	**i.** adrenergic
10. ___	cardiac plexus	**j.** white ramus
		k. stomach
		l. kidneys

Place in each blank the letter of the word that describes the change in activity produced by strongly stimulating the sympathetic or thoracolumbar division of the autonomic nervous system.

Column A	*Column B*
1. ___ blood vessels in skin	**a.** constricted
2. ___ blood vessels in abdominal viscera	**b.** dilated
3. ___ blood vessels in skeletal muscle	**c.** decreased
4. ___ blood pressure	**d.** increased
5. ___ blood sugar level	
6. ___ bronchioles of lung	
7. ___ flow of watery saliva	
8. ___ peristalsis of digestive tube	
9. ___ perspiration	
10. ___ pupil of eye	
11. ___ rate of heartbeat	
12. ___ rate of breathing	
13. ___ secretion of epinephrine	

C. True-False.   Place a *T* or *F* in the space provided.

___ **1.** The adrenal medulla is part of the sympathetic nervous system.

___ **2.** Stimulation of the parasympathetic system will cause the release of epinephrine from the adrenal medulla.

___ **3.** The primary function of the autonomic nervous system is the involuntary regulation of the internal environment.

___ **4.** In general, norepinephrine is released from postganglionic sympathetic nerve endings.

___ **5.** Acetylcholine is the transmitter agent released from postganglionic parasympathetic fibers.

___ **6.** All preganglionic sympathetic axons leave the spinal cord via the ventral root and then pass to a paravertebral ganglion via the white ramus.

___ **7.** The autonomic nervous system innervates skeletal muscles and glands.

___ **8.** Regulation of the autonomic nervous system occurs primarily in the medulla and hypothalamus of the brain.

___ **9.** A spinal nerve is formed by the junction of the dorsal and ventral roots.

___ **10.** Spinal shock may depress visceral reflexes controlling the emptying of the bladder and rectum.

___ **11.** The parasympathetic division of the autonomic nervous system consists of nerve fibers that arise from cells in the brain stem and sacral region of the spinal cord.

___ **12.** As a rule, the two divisions of the autonomic nervous system function in an antagonistic fashion.

___ **13.** Nerve impulse conduction can be compared to a self-propagating wave of negativity passing along a neuron's membrane.

___ **14.** Gray matter consists of a series of myelinated nerve fibers.

___ **15.** The hypothalamus has centers for initiating sympathetic and parasympathetic responses.

Answer Sheet—Chapter 9

Exercise 9.1

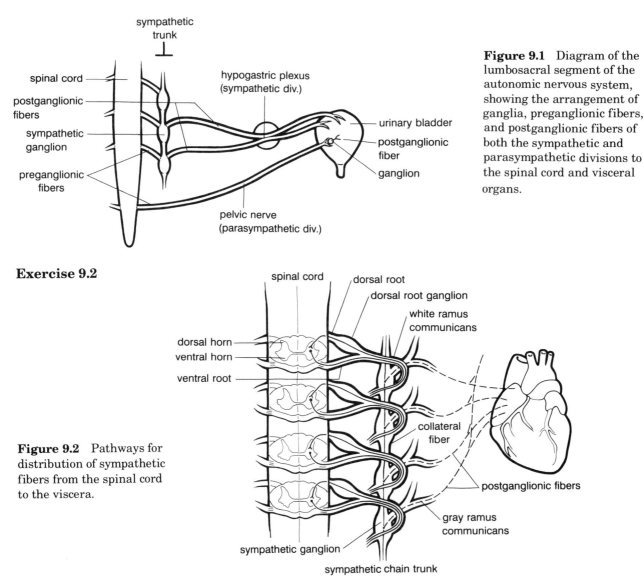

Figure 9.1 Diagram of the lumbosacral segment of the autonomic nervous system, showing the arrangement of ganglia, preganglionic fibers, and postganglionic fibers of both the sympathetic and parasympathetic divisions to the spinal cord and visceral organs.

Exercise 9.2

Figure 9.2 Pathways for distribution of sympathetic fibers from the spinal cord to the viscera.

Exercise 9.3

CHEMICAL TRANSMITTERS

Adrenergic Fibers

The terminal filaments of most sympathetic postganglionic neurons produce chiefly norepinephrine. This substance is also secreted by the medulla of the adrenal gland, and therefore these fibers are classified as adrenergic. Exceptions are the sympathetic fibers to sweat glands, to the blood vessels of the skin, and to the arrector muscles that elevate the hair. These postganglionic fibers enter the spinal nerves through the gray rami communicantes and reach the skin incorporated in peripheral nerves.

The effects of epinephrine, also secreted by the adrenal medulla, and norepinephrine can be widespread because these chemical substances, which result from excitation of the sympathetic postganglionic fibers, are carried by the bloodstream. Adrenergic transmitters stimulate the viscera controlled by the sympathetic division of the autonomic nervous system.

Cholinergic Fibers

Parasympathetic fibers also produce a chemical substance, acetylcholine, which is promptly converted to choline and acetic acid by the action of the enzyme cholinesterase. Acetylcholine is a depolarizing substance that aids synaptic transmission at the preganglionic terminals in both sympathetic and parasympathetic ganglia. At the postganglionic terminal of the parasympathetic or craniosacral division, secretion of acetylcholine acts to inhibit the organ. The ability of acetylcholine to excite at the first synapse and inhibit at the second synapse is due to differences in the chemistry of the target cells. In the heart, for example, parasympathetic stimulation is followed by hyperpolarization of the cardiac cells and inhibition of the heartbeat. The enzyme cholinesterase quickly destroys acetylcholine, which therefore has effects only locally, where it is secreted. Unlike norepinephrine, it probably is not carried by the bloodstream.

TEST ITEMS

A. 1.c, 2.c, 3.b, 4.a, 5.d, 6.b, 7.c, 8.b, 9.d, 10.c, 11.a, 12.a, 13.c, 14.d, 15.a, 16.c, 17.d, 18.a, 19.b, 20.d.

B. 1.j, 2.g, 3.i, 4.e, 5.h, 6.a, 7.c, 8.b, 9.k, 10.d.
    1.a, 2.a, 3.b, 4.d, 5.d, 6.b, 7.c, 8.c, 9.d, 10.b, 11.d, 12.d, 13.d.

C. 1.T, 2.F, 3.T, 4.T, 5.T, 6.T, 7.F, 8.T, 9.T, 10.T, 11.T, 12.T, 13.T, 14.F, 15.T.

Autonomic Nervous System

Across

1 a branch of a nerve or blood vessel

4 the middle region of an organ

5 to slow down or stop

6 a hormone that stimulates

9 pertaining to the vertebrae

12 group of nerve bodies

13 a group of nerves or blood vessels

14 pertaining to the outside

15 performing without free will

Down

1 enzyme that deactivates acetylcholine

2 chemicals that stimulate organs

3 self-controlling

7 pertaining to the intestines

8 below the thalmus

10 referring to the abdomen

11 chemicals that inhibit organs

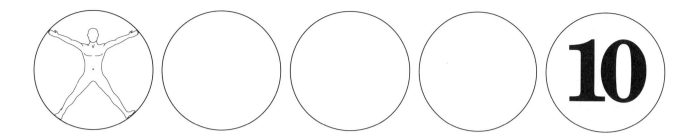

The Endocrine System

I. CHAPTER SYNOPSIS

The student is introduced to the principal organs of the endocrine system in terms of location, structure, hormones secreted and their physiological effects, and the disorders that result from abnormal secretions. Emphasis throughout is placed on the regulations of hormone secretions by negative feedback systems. There is also a detailed discussion of the general stress syndrome and the stages of the stress responses.

The endocrine and nervous systems are the major communications systems within the body. A hormone is a chemical substance synthesized by a specific organ (the endocrine gland) and secreted into the blood, which carries it to other sites where its action is exerted. The change in target tissue function resulting from the hormone's action is usually part of a negative feedback loop leading to the maintenance of the internal environment. Although each hormone circulates to all cells of the body, only certain cells are affected; this specificity depends on receptor sites to which the hormone attaches. The quantities of a hormone in the blood are determined by its rate of secretion, destruction, and excretion.

Most hormones cause changes in membrane transport or enzyme activity in the target cells; these primary effects can produce numerous secondary effects. The target cells for some hormones are themselves endocrine glands.

II. OBJECTIVES

After reading the chapter, the student should be able to:

- Explain the different mechanisms of hormone action.
- Explain the relationship between the central nervous system and endocrine glands.
- Explain the action of a hormone on a cell.
- Describe the feedback mechanism.
- Identify the endocrine glands and list their basic functions.

III. IMPORTANT TERMS

Using your textbook, define the following terms:

adenyl cyclase (ad'-en-il si'-klace) _____

_____

adrenal (ah-dreen'-ul) _____

_____

catecholamine (kat-ah-ko'-lah-meen) _____

_____

endocrine (en'-dah-krin) _____

_____

epinephrine (ep-ah-nef'-rin) _____

_____

feedback (feed'-bak) _____

_____

glycoprotein (gli-ko-pro'-teen) _____

_____

homeostasis (ho-mee-o-stay'-sis) _____

_____

hormone (hor'-mone) _____

_____

hypophysis (hi-pof'-ah-sis) _____

_____

hypothalamus (hi-po-thal'-ah-mus) _____

_____

insulin (in'-sah-lin) _____

_____

loop (loop) _____

_____

messenger (mes'-en-jur) _____

_____

neurosecretory (nyoor-o-seh-kreet'-ah-ree) _____

_____

ovaries (ov'-ah-rees) _____

_____

pancreas (pan'-kree-us) _____

_____

parathyroid (par-ah-thi'-roid) _____

_____

pituitary (pah-tew'-ah-ter-ee) _____

_____

prostaglandin (pros-tah-glan'-din) _____

_____

receptor (reh-sep'-tur) _____

_____

releasing factors (reh-lees'-ing fak'-turs) _____

_____

steroid (steer'-oid) _____

_____

target (tahr'-gut) _____

_____

testes (tes'-teez) _____

_____

thymus (thi'-mus) _____

_____

thyroid (thi'-roid) _____

_____

thyroxine (thi-rok'-sin) _____

_____

IV. EXERCISES

Complete the following exercises in the order given. A precise set of diagrams and terms has been chosen to describe the endocrine system.

Exercise 10.1

Labeling.   Write the name of the gland in the space provided. Color each gland differently.

Key:
adrenal
hypophysis
ovaries
pancreas
parathyroids
testes
thymus
thyroid

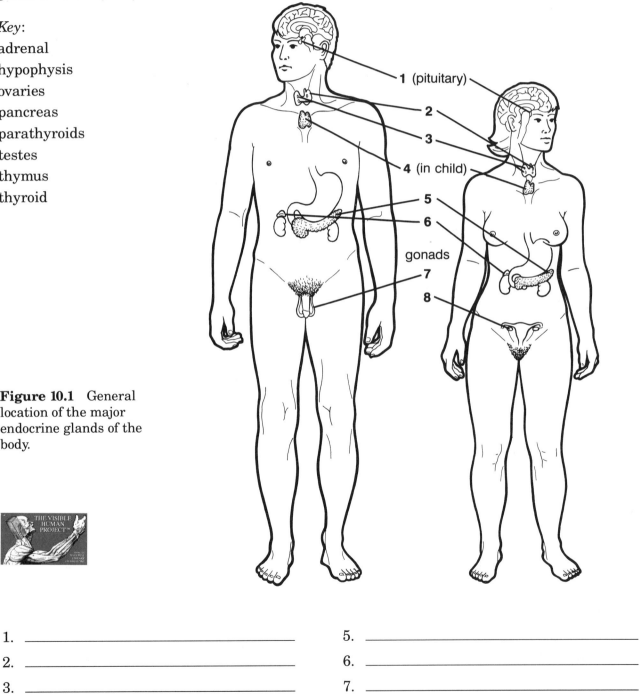

Figure 10.1   General location of the major endocrine glands of the body.

1. _____

2. _____

3. _____

4. _____

5. _____

6. _____

7. _____

8. _____

Exercise 10.2

Completion.   Write the term in the space provided.

Key:

activities

another

blood

carried

effective

effects

glycoproteins

homeostasis

hormones

inhibit

maintaining

messengers

one

organic

proteins

steroids

stimulate

target

HORMONES

Chemical Nature of Hormones

_____ are _____ compounds of varying structural complexity that are _____ by the _____ to other parts of the body where they exert their specific _____ . They are produced in _____ gland or part of a gland and are_____ in _____ gland or in another region of that same gland. In simple terms, hormones are chemical _____ that pass via the bloodstream to the _____ organ or process. They may either _____ or _____ a function, but in general they do not initiate a process. Hormones are either _____ , _____ (combination of a protein and a sugar), or _____ (substance made of a fat-soluble carbon). The one thing that all hormones have in common, whether protein, glycoprotein, or steroid, is the function of _____ _____ by modifying the physiological _____ of cells.

Key:

activates

adenyl cyclase

AMP

ATP

beat

binds

catecholamine

effect

few

increase

influence

interacting

life

metabolism

prostaglandins

receptors

regulate

responses

secondary

sites

small

stimulation

target

Actions of Hormones

Hormones are effective in remarkably _____ quantities. For example, the injection of a _____ micrograms of epinephrine into a dog causes a definite _____ in the rate of heart_____ . There is little doubt that each hormone has some _____ on the fundamental _____ of its target cells or tissues. Hormones, therefore, have a marked _____ on such basic _____ processes as growth, development, reproduction, energy utilization, and cell permeability.

Much is known about the way hormones work on their target tissues or cells. Protein and _____ (steroid-stimulating) hormones act by first _____ with receptor _____ on the cell membrane. The cell membrane contains the _____ system. The hormone _____ to specific _____ in the cell membrane and subsequently _____ adenyl cyclase. This enzyme converts adenosine triphosphate (_____) into 3, 5-cyclic _____ , which acts as the secondary messenger. Cyclic AMP, the _____ messenger, then moves to other structures.

Although it is not known how the group of hormones called _____ directly acts upon cells, it is postulated that they somehow _____ the formation of cyclic AMP and may be involved in the _____ of _____ tissues to hormonal _____ .

Exercise 10.3

Labeling. Write the name of the term or structure in the space provided. Color the separate messengers differently.

Key:

adenyl
AMP
ATP
endocrine
first
membrane
messenger
neurosecretory
prostaglandin
second
specialized

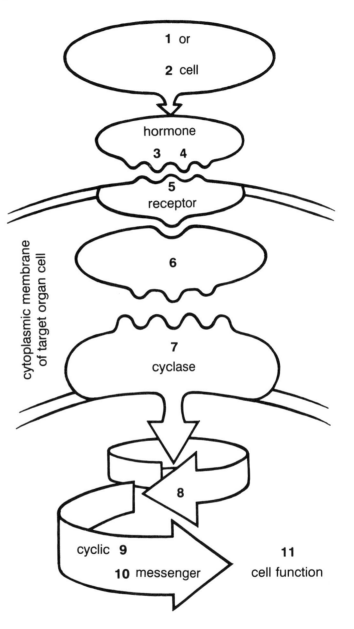

Figure 10.3 Mechanism of hormone action. Hormones act as messengers delivering their message to a membrane receptor in the target cell. Prostaglandins appear to regulate hormonal activity by influencing adenyl cyclase and cyclic AMP activity within the cell. Cyclic AMP serves as the second messenger for specialized cell functions.

1. _____

2. _____

3. _____

4. _____

5. _____

6. _____

7. _____

8. _____

9. _____

10. _____

11. _____

Exercise 10.4

Completion.   Write the term in the space provided.

Key:

common	regulate
continually	regulated
directly	release
hypothalamus	releasing
indirectly	role
minimal	specific
needs	stored
nervous	

Feedback Control Systems

Hormones are secreted _____, and their rate of secretion is _____ by the demands of the body _____ . The _____ system controls the endocrine system either _____ or _____. The direct influence is _____ and best illustrated by the effects of the sympathetic nervous system on the secretion of the adrenal medulla. Indirect control is most _____ and is centered around the _____ of the _____. The hypothalamus secretes certain hormones and transfers these hormones to the posterior pituitary where they are _____ . Also, the hypothalamus secretes chemical substances, known as _____ factors, that are released into the vascular bed between the hypothalamus and the anterior pituitary. These releasing factors are _____ and _____ the _____ of the anterior pituitary hormones.

Key:

effect	negative
feed back	positive
hormone	releasing
inhibit	stimulate
inhibiting	target

The _____ released by the _____ endocrine gland may _____ and influence release of the _____ or _____ substance of the hypothalamus or the pituitary hormone. If the _____ of the feedback is to _____ the overall response of the system, it is termed _____ feedback. If the effect of the feedback is to _____ further hormone release, it is _____ feedback.

Key:

back
direct
high
influence
inhibit
inhibits
long feedback
mechanisms
regulate
release
secretion
short feedback
thyroxine

In many cases the _____ of a hormone is regulated by both positive and negative feedback _____ . _____ , which is released by the thyroid gland, can _____ the release of thyroid-stimulating hormone in two ways. First, _____ levels _____ the release of thyrotropin-releasing hormone by the hypothalamus, which reduces TSH release by the anterior pituitary. Second, thyroxine _____ the _____ of TSH by _____ effect on the cells of the anterior pituitary. This particular type of feedback pathway (from target gland to pituitary brain) is called a _____ loop. Additionally, the pituitary hormones may feed _____ to the brain to ___ _____ the releasing or inhibiting substances; this pathway is called a _____ loop.

Exercise 10.5

Labeling. Write the name of the term or structure in the space provided. Color the feedback loops differently to the anatomy.

Key:
central
feedback
hormone
hypophyseal
long
loop
neurosecretory
releasing
short
stimuli
target

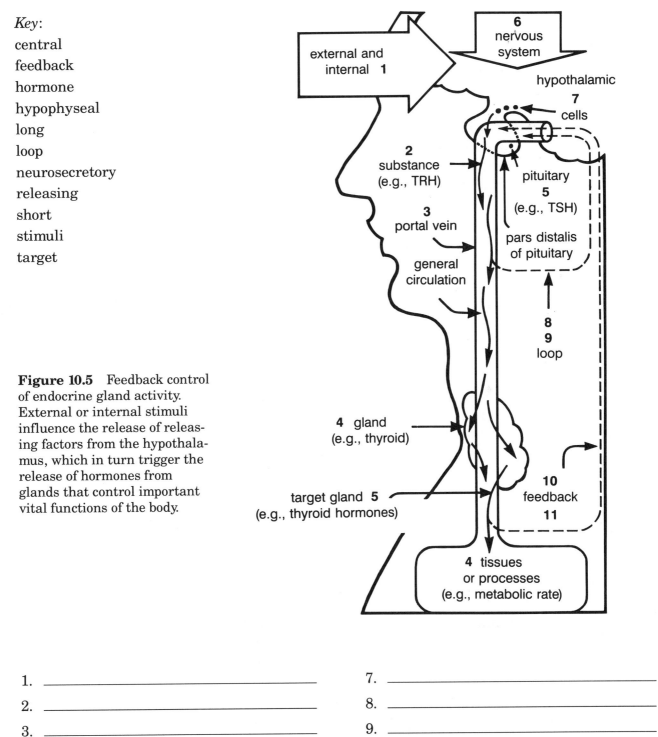

Figure 10.5 Feedback control of endocrine gland activity. External or internal stimuli influence the release of releasing factors from the hypothalamus, which in turn trigger the release of hormones from glands that control important vital functions of the body.

1. _____   7. _____

2. _____   8. _____

3. _____   9. _____

4. _____   10. _____

5. _____   11. _____

6. _____

V. TEST ITEMS

A. *Multiple Choice.* There is only one answer that is either correct or most appropriate. Circle the answer that corresponds to the question.

1. Insulin is secreted by the pancreas when
   a. blood sugar level is high.
   b. blood sugar level is low.
   c. glucagon is secreted.
   d. the gallbladder secretes bile.

2. In a feedback system,
   a. input is fed back into the system.
   b. input serves no useful purpose.
   c. output is fed back into the system.
   d. output is never an important factor.

3. Hormones
   a. are produced by exocrine glands.
   b. are carried in the blood to virtually all parts of the body.
   c. remain at constant concentration in the blood.
   d. none of the above

4. Which endocrine gland is not under the control of the anterior pituitary gland?
   a. the thyroid
   b. the parathyroids
   c. adrenal cortex
   d. the gonads
   e. the corpus luteum

5. During general stress syndrome, the alarm reaction is initiated by
   a. neurohumor of the hypothalamus.
   b. hormones of the adrenal cortex.
   c. hypothalamic stimulation of the sympathetic nervous system and adrenal medulla.
   d. hormones of the anterior pituitary.

6. A child who exhibits dwarfism, mental retardation, yellowish skin color, and fat pads in the face and abdomen is probably suffering from
   a. cretinism.
   b. pituitary dwarfism.
   c. acromegaly.
   d. goiter.

7. An individual with spasms, twitches, and convulsions probably has a defective
   a. thyroid.
   b. parathyroid.
   c. adrenal medulla.
   d. pineal.

8. The fight-or-flight responses that are brought about by sympathetic stimulation are duplicated by the action of which endocrine gland?
   a. anterior pituitary
   b. adrenal medulla
   c. thyroid
   d. posterior pituitary

9. In a strict sense, the posterior pituitary is not an endocrine gland because it
   a. has a rich blood supply.
   b. does not make hormones.
   c. is not near the brain.
   d. contains ducts.

10. Which gland is primarily concerned with sodium and potassium salt balance?
    a. parathyroid
    b. thyroid
    c. adrenal cortex
    d. anterior pituitary

11. Which hormone increases the reabsorption of sodium by kidney tubules?
    a. parathormone
    b. thyroxine
    c. oxytocin
    d. aldosterone

12. Hyperglycemia, increased urine production, thirst, and ketosis are all symptoms of
    a. hyperthyroidism.
    b. diabetes mellitus.
    c. diabetes insipidus.
    d. hyperinsulinism.

13. If a person is diagnosed as having a high metabolic rate, which endocrine gland is probably malfunctioning?
    a. parathyroid
    b. thymus
    c. posterior pituitary
    d. thyroid

14. Water retention that may lead to edema is related to a hypersecretion of
    a. aldosterone.
    b. cortisone.
    c. thyroxin.
    d. follicle-stimulating hormone.

15. Removal of the anterior pituitary would effect all but the
    a. mammary glands.
    b. adrenal cortex.
    c. thyroid gland.
    d. adrenal medulla.

16. The somatotropic hormone does which of the following?
    a. regulates the growth of the skeleton
    b. helps to regulate the activities of the adrenal glands
    c. causes lymphocytes to be produced
    d. causes decreased activity of the exocrine portion of the pancreas

17. Parathormone does which of the following?
    a. It aids in the maintenance of normal blood levels of calcium and phosphorus.
    b. It aids in the maintenance of normal blood levels of sodium and potassium.
    c. It influences the production of prolactin.
    d. It responds to the stimulus of the thyrotropic hormone.

18. The function of calcitonin is to
    a. prevent calculi from forming.
    b. increase blood sodium levels.
    c. lower blood calcium levels.
    d. cause calcification of the pineal gland at the time of puberty.

19. Proper carbohydrate metabolism depends on the production of insulin by the
    a. parathyroid.   c. liver.
    b. adrenal.   d. pancreas.

20. Insulin
    a. is a steroid hormone.
    b. is secreted by cells in the pancreas in response to decreased blood glucose levels.
    c. increases glycogen synthesis by increasing glucose concentration within the cells and by increasing enzyme activity.
    d. increases the blood concentration of glucose.

B. *Matching Questions.*   Each of the phrases in COLUMN B refers to a word or phrase in COLUMN A. Insert the letter of the word or phrase from COLUMN B that best describes it. Some words or phrases may be used more than once or not at all.

Column A	*Column B*
1. ___ adrenal medulla	**a.** iodine
2. ___ stimulates the ovary	**b.** aldosterone
3. ___ regulates Na^+ and K^+	**c.** epinephrine
4. ___ contraction of smooth muscle	**d.** vasopressin
5. ___ controls body growth	**e.** progesterone
6. ___ male sex characteristics	**f.** testosterone
7. ___ another form of estrogen	**g.** estradiol
8. ___ regulates Ca^{2+} metabolism	**h.** parathyroid hormone
9. ___ inhibits ovarian function	**i.** insulin
10. ___ stimulates pancreas	**j.** norepinephrine
11. ___ decrease blood sugar	**k.** glucagon
12. ___ thyroid gland	**l.** growth hormone
13. ___ alpha cells of the pancreas	**m.** prolactin
14. ___ stimulates uterine muscle	**n.** oxytocin
15. ___ stimulates breakdown of liver glycogen	**o.** secretin
	p. melatonin

C. *True-False.*   Write a *T* or *F* in the space provided.

____ **1.** People with kidney or liver disease may suffer from hormone excesses due to a decreased rate of destruction of the hormones by these organs.

____ **2.** One of the ways hormones work is to alter the transport of certain molecules across cell membranes.

____ **3.** Some reflexes are mediated by hormones.

____ **4.** Hormones are secreted by exocrine glands into ducts leading to the body cavities.

____ **5.** Hormones must be present in high concentrations in the blood in order to have any physiological effect on the organism.

____ **6.** In general, the end result of most hormone action is a change in the metabolic activity in the target-organ cells.

____ **7.** Hormones may be released from the terminals of neurons.

____ **8.** The hypothalamic-releasing factors are not true hormones because they do not reach their target tissue by way of the circulatory system.

____ **9.** The release of a hormone into the blood may be initiated in some cases by stimulating the nerves into the gland.

____ **10.** Hormones cannot cause their target cells to perform new activities but only accelerate or modify the existing capabilities of the cell.

____ **11.** Hormones are necessary for the regulation of the internal environment under the varying conditions imposed on the organism by a changing external environment.

____ **12.** Endocrine glands use ducts whereas exocrine glands are ductless.

____ **13.** Long-distance communication between cells is accomplished by nerves and/or hormones.

____ **14.** A type of feedback in which an increase in the output of a system results in a decrease in the input is known as positive feedback.

____ **15.** Releasing factors of the hypothalamus stimulate the release by the adenohypophysis of TSH, ACTH, LH, and growth hormone.

____ **16.** That portion of the adrenal gland essential for life is the cortex.

____ **17.** Negative feedback results in an increase in the production of hormones by the hypophysis.

____ **18.** The pancreas is both an exocrine and endocrine gland.

____ **19.** Thyroid-stimulating hormone is a trophic hormone.

____ **20.** Hormones are secreted by their ducts to their target organ(s).

Answer Sheet—Chapter 10

Exercise 10.1

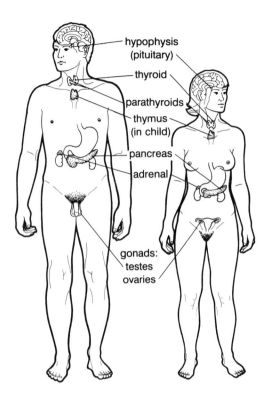

hypophysis (pituitary)
thyroid
parathyroids
thymus (in child)
pancreas
adrenal
gonads: testes ovaries

Exercise 10.2

HORMONES

Chemical Nature of Hormones

<u>Hormones</u> are <u>organic</u> compounds of varying structural complexity that are <u>carried</u> by the <u>blood</u> to other parts of the body where they exert their specific <u>effects</u>. They are produced in <u>one</u> gland or part of a gland and are <u>effective</u> in <u>another</u> gland or in another region of that same gland. In simple terms, hormones are chemical <u>messengers</u> that pass via the bloodstream to the <u>target</u> organ or process. They may either <u>stimulate</u> or <u>inhibit</u> a function, but in general they do not initiate a process. Hormones are either <u>proteins</u>, <u>glycoproteins</u> (combination of a protein and a sugar), or <u>steroids</u> (substance made of a fat-soluble carbon). The one thing that all hormones have in common, whether protein, glycoprotein, or steroid, is the function of <u>maintaining</u> <u>homeostasis</u> by modifying the physiological <u>activities</u> of cells.

Actions of Hormones

Hormones are effective in remarkably <u>small</u> quantities. For example, the injection of a <u>few</u> micrograms of epinephrine into a dog causes a definite <u>increase</u> in the rate of heart <u>beat</u>. There is little doubt that each hormone has some <u>effect</u> on the fundamental <u>metabolism</u> of its target cells or tissues. Hormones, therefore, have a marked <u>influence</u> on such basic <u>life</u> processes as growth, development, reproduction, energy utilization, and cell permeability.

Much is known about the way hormones work on their target tissues or cells. Protein and <u>catecholamine</u> (steroid-stimulating) hormones act by first <u>interacting</u> with receptor <u>sites</u> on the cell membrane. The cell membrane contains the <u>adenyl cyclase</u> system. The hormone <u>binds</u> to specific <u>receptors</u> in the cell membrane and subsequently <u>activates</u> adenyl cyclase. This enzyme converts adenosine triphosphate (<u>ATP</u>) into 3,5-cyclic <u>AMP</u>, which acts as the secondary messenger. Cyclic AMP, the <u>secondary</u> messenger, then moves to other structures.

Although it is not known how the group of hormones called <u>prostaglandins</u> directly acts upon cells, it is postulated that they somehow <u>regulate</u> the formation of cyclic AMP and may be involved in the <u>responses</u> of <u>target</u> tissues to hormonal <u>stimulation</u>.

Exercise 10.3

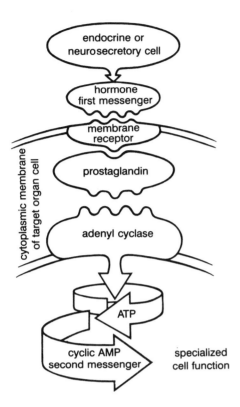

Figure 10.3 Mechanism of hormone action. Hormones act as messengers delivering their message to a membrane receptor in the target cell. Prostaglandins appear to regulate hormonal activity by influencing adenyl cyclase and cyclic AMP activity within the cell. Cyclic AMP serves as the second messenger for specialized cell functions.

Exercise 10.4

Feedback Control Systems

Hormones are secreted continually, and their rate of secretion is regulated by the demands of the body needs. The nervous system controls the endocrine system either directly or indirectly. The direct influence is minimal and best illustrated by the effects of the sympathetic nervous system on the secretion of the adrenal medulla. Indirect control is most common and is centered around the role of the hypothalamus. The hypothalamus secretes certain hormones and transfers these hormones to the posterior pituitary where they are stored. Also, the hypothalamus secretes chemical substances, known as releasing factors, that are released into the vascular bed between the hypothalamus and the anterior pituitary. These releasing factors are specific and regulate the release of the anterior pituitary hormones.

The hormone released by the target endocrine gland may feed back and influence release of the releasing or inhibiting substance of the hypothalamus or the pituitary hormone. If the effect of the feedback is to inhibit the overall response of the system, it is termed negative feedback. If the effect of the feedback is to stimulate further hormone release, it is positive feedback.

In many cases the secretion of a hormone is regulated by both positive and negative feedback mechanisms. Thyroxine, which is released by the thyroid gland, can regulate the release of thyroid-stimulating hormone in two ways. First, high levels inhibit the release of thyrotropin-releasing hormone by the hypothalamus, which reduces TSH release by the anterior pituitary. Second, thyroxine inhibits the release of TSH by direct effect on the cells of the anterior pituitary. This particular type of feedback pathway (from target gland to pituitary brain) is called a long feedback loop. Additionally, the pituitary hormones may feed back to the brain to influence the releasing or inhibiting substances; this pathway is called a short feedback loop.

Exercise 10.5

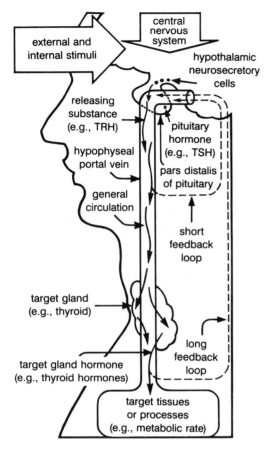
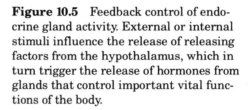

Figure 10.5 Feedback control of endocrine gland activity. External or internal stimuli influence the release of releasing factors from the hypothalamus, which in turn trigger the release of hormones from glands that control important vital functions of the body.

Test Items

A. 1.a, 2.c, 3.b, 4.b, 5.c, 6.a, 7.b, 8.b, 9.b, 10.c, 11.d, 12.b, 13.d, 14.a, 15.d, 16.a, 17.a, 18.c, 19.d, 20.c.

B. 1.j, 2.e, 3.b, 4.d, 5.l, 6.f, 7.g, 8.h, 9.p, 10.o, 11.i, 12.a, 13.k, 14.n, 15.c.

C. 1.T, 2.T, 3.T, 4.F, 5.F, 6.T, 7.T, 8.F, 9.T, 10.T, 11.T, 12.F, 13.T, 14.F, 15.T, 16.T, 17.F, 18.T, 19.T, 20.F.

Endocrine System

Across

1 organic substance produced by the adrenal gland

4 gland attached to the kidney

6 a dynamic equilibrium

11 a stimulant called adrenaline

14 a carbohydrate-protein found in the cell membrane

15 the pituitary gland

16 a carrier molecule to a cell

Down

2 area of reaction on a cell membrane

3 amine compounds that stimulate

5 ductless glands with internal secretions

7 hormone that regulates carbohydrate metabolism

8 area below the thalamus in the brain

9 hormone from the thyroid that controls metabolism

10 control mechanism to regulate hormone secretion

12 the specific tissue of hormonal action

13 chemical produced in one gland that is effective in another area or gland

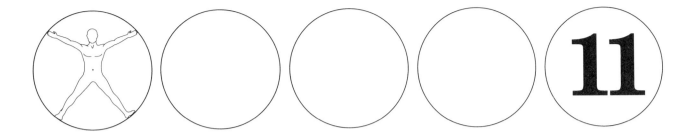

The Special Senses

I. CHAPTER SYNOPSIS

Our perception of the external world is determined by the physiologic mechanisms involved in the processing of sensory information transmitted by nerve fibers to the brain from various receptors located throughout the body.

Sensations may be classified as cutaneous, visceral, olfactory, gustatory, visual, auditory, and position sense. The three components of a sensory mechanism are a sense organ or receptor, a pathway to the brain, and a sensory area in the cerebral cortex. In a sense organ or receptor the property of excitability is most highly developed, and each is specialized to respond to a specific type of stimulus. The process of projection, the sensation of afterimages, adaptation, and variability of intensity are all characteristics of sensations. Touch, pressure, heat, cold, and pain comprise the cutaneous senses. There are two types of pain: visceral and somatic. Somatic pain has been subdivided into superficial or cutaneous pain and deep pain. Muscle sense, also called proprioceptive or kinesthetic sense, provides nonvisual information on the position of body parts. Organic sensation and visceral pain impulses are initiated by visceral receptors. The olfactory epithelium of the nasal cavity and the taste buds in the papillae of the tongue are the location of receptors for the senses of smell and taste. Receptors for vision lie in the retina of the eyeball and are discussed with associated structures of the eye that make vision possible. Hearing, the sense by which sounds are appreciated, involves the function of structures of the external ear, middle ear, and cochlear portions of the inner ear.

Position sense involves the orientation of the head in space, the movement of the body through space, and the sense of balance and equilibrium of the body. The vestibule and semicircular canals of the inner ear are used in position sense.

II. OBJECTIVES

After reading the chapter, the student should be able to:

- Describe the sensory mechanisms.
- Differentiate among interceptors, proprioceptors, and exteroceptors, and give examples of each.
- Define chemoreceptor, pressoreceptor, and photoreceptor and relate their significance.
- Describe referred pain.
- Describe the anatomy of the eyeball and its protective structures.
- Explain the physical phenomenon of refraction and how it operates in focusing.
- Explain depth perception and relate binocular vision with diplopia and hemiopia.
- Describe the anatomy of the ear.
- Follow the transmission of sound from the tympanic membrane to the basilar membrane.
- Explain the Place theory of hearing.
- Differentiate between static and dynamic equilibrium.

III. IMPORTANT TERMS

Using your textbook, define the following terms:

accommodation (ah-kom-ah-day′-shun) _____

_____

adaptation (ad-ap-tay′-shun) _____

_____

astigmatism (ah-stig′-mah-tiz-em) _____

_____

auditory (awd′-ah-tor-ee) _____

_____

auricle (awr'-i-kul) _____

_____

cataract (kat'-ah-rakt) _____

_____

chemoreceptor (kee'-mo-reh-sep-tur) _____

_____

choroid (kor'-oid) _____

_____

cochlea (kok'-lee-ah) _____

_____

concave (kon'-kave) _____

_____

conjunctivitis (kahn-junk-ti-vite'-us) _____

_____

convex (kon'-veks) _____

_____

cornea (kor'-nee-ah) _____

_____

Corti (kor'-tee) _____

_____

diplopia (dip-lo'-pee-ah) _____

_____

Eustachian (yoo-stay'-shun) _____

_____

exteroceptors (ek-stah-ro-sep'-turs) _____

_____

fovea (fo'-vee-ah) _____

_____

glaucoma (glaw-ko'-mah) _____

_____

hyperopia (hi-pah-ro'-pee-ah) _____

_____

incus (ing'-kus) _____

_____

interoceptors (int-ah-ro-sep'-turs) _____

_____

iris (i'-rus) _____

_____

malleus (mal'-ee-us) _____

_____

myopia (mi-o'-pee-ah) _____

_____

nystagmus (nis-tag'-mus) _____

_____

optic (op'-tik) _____

_____

photoreceptor (fote-o-reh-sep'-tur) _____

_____

pupil (pew'-pil) _____

_____

refraction (re-frak'-shun) _____

_____

retina (ret'-in-ah) _____

_____

rhodopsin (ro-dop'-sin) _____

_____

sclera (skler'-ah) _____

_____

stapes (stay'-peez) _____

_____

tympanic (tim-pan'-ik) _____

_____

vestibular (ve-stib'-yah-lur) _____

_____

IV. EXERCISES

Complete the following exercises in the order given. A precise set of terms
and diagrams has been chosen to describe the special senses.

Exercise 11.1

Completion.   Write the term in the space provided.

Key:

alike

brain

change

CNS

composition

conscious

decodes

dendrites

environments

impulses

information

intensity

interpret

interpreted

interprets

odor

other

pathways

peripheral

photoreceptors

pigments

provide

receptor

receptors

sensation

sense

specialized

stimuli

structure

tissue

unconscious

SENSORY MECHANISMS

A _____ organ, or _____ , is a
_____ nervous _____ situated at the
_____ endings of the _____ of afferent
neurons. The receptor's primary function is to _____
the body with _____, both _____
and _____ , about degrees of _____ in the
organism's external and internal _____ .

Three important features of sense organs should be emphasized:

1. Specific _____ are particularly sensitive to
   specific _____ . However they can also respond
   to _____ stimuli of sufficient _____.
   For example, pressure on the eyeball causes a sensation of
   light.

2. Specific sensitivity to certain stimuli is due to the
   _____ and _____ of the receptor.
   For example, light-absorbing _____
   are found in the _____ of the eye.

3. The type of _____ elicited by a receptor depends
   on which nerve _____ are activated, not on how
   they are activated. All nerve _____ are
   essentially _____, regardless of the
   stimulus that initiates them. The impulse is transmitted to
   the _____ where it is _____ . For
   example, the region of the cerebrum receiving impulses
   from an olfactory receptor _____ and _____
   them as a specific odor or aroma. If an olfactory receptor
   is artificially stimulated, the same nerve pathways to the
   _____ will be activated, and the brain will
   _____ the arriving impulses as a specific
   _____.

Exercise 11.2

Labeling.   Write the name of the structure in the space provided. Color the three layers of the eye differently.

Key:

capsule	conjunctiva	iris	optic	rectus	superior
choroid	cornea	lens	posterior	retina	vision
ciliary	fovea	ligaments	pupil	sclera	

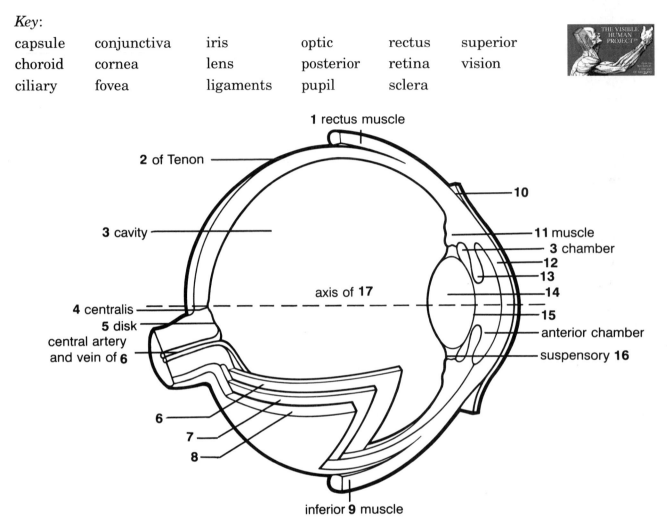

1 rectus muscle

2 of Tenon

10

3 cavity

11 muscle
3 chamber
12
13

axis of **17**

14

4 centralis
5 disk
central artery
and vein of **6**

15

anterior chamber

suspensory **16**

6
7
8

inferior **9** muscle

Figure 11.2   Structure of the eye, transverse section.

1. _____   10. _____

2. _____   11. _____

3. _____   12. _____

4. _____   13. _____

5. _____   14. _____

6. _____   15. _____

7. _____   16. _____

8. _____   17. _____

9. _____

Exercise 11.3

Labeling.   Write the name of the structure in the space provided. Color the temporal bone differently from the ear mechanism.

Key:

auditory	Corti	inner	middle	stapes
auricle	Eustachian	jugular	oval	temporal
cochlea	external	malleus	round	tympanic
cochlear	incus	meatus	semicircular canals	vestibular

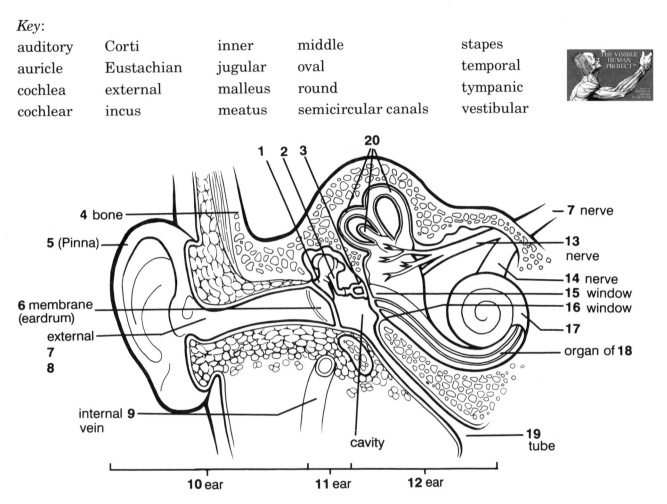

Figure 11.3   Frontal diagram of the outer ear, middle ear, and internal ear. A section of the cochlear duct has been cut away to show the position of the organ of Corti.

1. _____
2. _____
3. _____
4. _____
5. _____
6. _____
7. _____
8. _____
9. _____
10. _____

11. _____
12. _____
13. _____
14. _____
15. _____
16. _____
17. _____
18. _____
19. _____
20. _____

V. TEST ITEMS

A. *Multiple Choice.* There is only one answer that is either correct or most appropriate. Circle the answer that corresponds to the question.

1. As a result of nerve damage, a person cannot see at all with the left eye but has no trouble seeing with the right eye. The site of injury is probably in the
   a. optic chiasma.
   b. left occipital lobe.
   c. left optic tract.
   d. left optic nerve.

2. A physician who specializes in disorders of the eye is a(n)
   a. optician.
   b. optometrist.
   c. ophthalmologist.
   d. neurologist.

3. Otitis media may follow throat infections because the
   a. auditory canal connects the middle ear with the tympanic membrane.
   b. Eustachian tube opens into the inner ear.
   c. mucosa of the nasopharynx is continuous with the mucosa of the middle ear.
   d. mastoid sinus opens into the middle ear.

4. Which of the following is likely to result in conduction deafness?
   a. stiffness of the joints between the bones of the middle ear
   b. hardening of the perilymph
   c. loss of the organs of Corti
   d. sclerosis of the membranous labyrinth

5. The blind spot is the place where
   a. there are more rods than cones.
   b. there are more cones than rods.
   c. the optic nerve leaves the eyeball.
   d. the iris attaches to the cornea.

6. The strong white fibrous outer coat of the eye is called the
   a. cornea.
   b. choroid.
   c. sclera.
   d. lens.

7. The organs of Corti are receptors for
   a. light rays.
   b. equilibrium.
   c. sound waves.
   d. taste.
   e. smell.

8. Aqueous humor drains from the anterior chamber into the
   a. vitreous body.
   b. lacrimal duct.
   c. mastoid sinuses.
   d. canal of Schlemm.

9. Receptors that are stimulated by activities that occur in muscles and articulations are termed
   a. exteroceptors.
   b. interoceptors.
   c. mechanoreceptor.
   d. proprioceptors.

10. Sound waves are translated into nerve impulses in the
    a. tympanic membrane.   c. cochlea.
    b. semicircular canals.   d. fovea centralis.

11. The organs of Corti
    a. are concerned with equilibrium.
    b. influence station and gait.
    c. change vibrations into nerve impulses.
    d. give rise to the vestibular nerve.

12. The fovea centralis is the place where
    a. nerve impulses for equilibrium are concentrated.
    b. visual acuity is greatest.
    c. high-pitched tones are received.
    d. the optic nerve leaves the eyeball.

13. The fluid that fills the membranous labyrinth is called
    a. aqueous humor.   c. endolymph.
    b. vitreous humor.   d. perilymph.

14. The chemical reaction associated with vision takes place in the
    a. aqueous humor.   c. vitreous humor.
    b. iris.   d. retina.

15. Which of these is a true statement?
    a. Hearing does not depend on the inner ear.
    b. All parts of the organ of Corti hear all ranges of sound.
    c. The loud music young people listen to cannot damage their ears.
    d. Hearing depends on pressure waves.

16. Puncturing the eardrum so that it is inoperative would
    a. not affect your sense of hearing.
    b. prevent the normal transmission of sound vibration.
    c. destroy the sense receptors for hearing.
    d. account for why some people who hear still cannot sing on tune.

17. The cochlea
    a. is a coiled structure found in the inner ear.
    b. contains three fluid-filled canals.
    c. contains the organ of Corti.
    d. all of the above

18. The bones of the middle ear
    a. respond to a change in the position of the head.
    b. transmit sound waves.
    c. are sense receptors connected to the auditory nerve.
    d. all of the above

19. A mucous membrane that lines the inner surface of the eyelid and continues as the surface layer of the eyeball is called
    a. the conjunctiva.
    b. the choroid.
    c. the sclera.
    d. none of the above

20. The senses of taste and smell are detected by
   a. thermoreceptors.
   b. chemoreceptors.
   c. electromagnetic receptors.
   d. mechanoreceptors.

B. *Matching Questions.* Each of the phrases in COLUMN B refers to a word or phrase in COLUMN A. Insert the letter of the word or phrase from COLUMN B that best describes it. Some words or phrases may be used more than once or not at all.

Column A	*Column B*
1. ___ sensory unit	**a.** sensitive to pressure changes
2. ___ interoceptors	**b.** the distance between two successive wave peaks
3. ___ lacrimal gland	**c.** associated with pain sensitivity
4. ___ wavelength	**d.** opaque lens cells
5. ___ modalities	**e.** receptors located in the walls of the viscera
6. ___ cataract	**f.** nearsighted
7. ___ free nerve endings	**g.** changes the shape of the lens
8. ___ myopic eye	**h.** keeps the eyeball moist
9. ___ ciliary muscles	**i.** a single afferent neuron plus all the receptors it innervates make up a
10. ___ mechanoreceptor	**j.** therapy

Column A	*Column B*
1. ___ hyperopic	**a.** a protein
2. ___ opsin	**b.** anvil
3. ___ optic chiasma	**c.** optic crossing
4. ___ incus	**d.** detects changes in both motion and posture
5. ___ cochlea	**e.** related to the loudness of sound
6. ___ vestibular system	**f.** inner ear
7. ___ amplitude	**g.** elevates the eye
8. ___ superior rectus	**h.** a farsighted eye
9. ___ inferior rectus	**i.** depresses the eye
10. ___ erythrolable	**j.** a photopigment

C. *True-False.*   Write a *T* or *F* in the space provided.

_____ **1.** The crystalline lens has a fixed refractive index.

_____ **2.** The aqueous humor of the eye, like cerebrospinal fluid, is constantly being manufactured and reabsorbed.

_____ **3.** The iris of the eye and ciliary body constitute the extrinsic muscles of the eye.

_____ **4.** The near point of vision moves closer with increasing age.

_____ **5.** Pain frequently seems to arise at locations other than the area in which a disorder occurs.

_____ **6.** Adaptation to odors is a relatively slow process.

_____ **7.** The sensory components of cranial nerves conveying impulses from the special receptors have their cell bodies outside the brain.

_____ **8.** All taste sensation depends upon the integrity of the seventh cranial nerve.

_____ **9.** Hair cells in contact with the tectorial membrane are stimulated when the membrane vibrates.

_____ **10.** Deafness due to injury to the auditory center of the brain could be alleviated by the use of a hearing aid.

_____ **11.** Otoliths are associated with body equilibrium.

_____ **12.** The inner ear is characterized by the presence of auditory ossicles.

_____ **13.** Impulses associated with hearing pass from the utricle and saccule to the vestibulocochlear nerve.

_____ **14.** Pressure within the ear is relieved by the passage of air through the Eustachian tube.

_____ **15.** The part of the body affected by impacted cerumen is the anterior chamber of the eye.

_____ **16.** The macula lutea contains the area of sharpest vision.

_____ **17.** The vitreous humor is situated between the lens and retina.

_____ **18.** Taste and smell are perceived through stimulation of dissolved substances acting upon chemoreceptors (chemical receptors).

_____ **19.** The stapes fits into the fenestra cochlea or round window.

_____ **20.** The portion of the basilar membrane involved increases as sound intensity is increased.

Answer Sheet—Chapter 11

Exercise 11.1

SENSORY MECHANISMS

A <u>sense</u> organ, or <u>receptor</u>, is a <u>specialized</u> nervous <u>tissue</u> situated at the <u>peripheral</u> endings of the <u>dendrites</u> of afferent neurons. The receptor's primary function is to <u>provide</u> the body with <u>information</u>, both <u>conscious</u> and <u>unconscious</u>, about degrees of <u>change</u> in the organism's external and internal <u>environments</u>.

Three important features of sense organs should be emphasized:

1. Specific <u>receptors</u> are particularly sensitive to specific <u>stimuli</u>. However, they can also respond to <u>other</u> stimuli of sufficient <u>intensity</u>. For example, pressure on the eyeball causes a sensation of light.

2. Specific sensitivity to certain stimuli is due to the <u>structure</u> and <u>composition</u> of the receptor. For example, light-absorbing <u>pigments</u> are found in the <u>photoreceptors</u> of the eye.

3. The type of <u>sensation</u> elicited by a receptor depends on which nerve <u>pathways</u> are activated, not on how they are activated. All nerve <u>impulses</u> are essentially <u>alike</u>, regardless of the stimulus that initiates them. The impulse is transmitted to the <u>CNS</u> where it is <u>interpreted</u>. For example, the region of the cerebrum receiving impulses from an olfactory receptor <u>decodes</u> and <u>interprets</u> them as a specific odor or aroma. If an olfactory receptor is artificially stimulated, the same nerve pathways to the <u>brain</u> will be activated, and the brain will <u>interpret</u> the arriving impulses as a specific <u>odor</u>.

Exercise 11.2

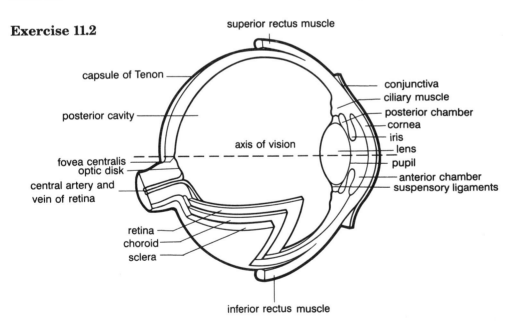

Figure 11.2   Structure of the eye, transverse section.

Exercise 11.3

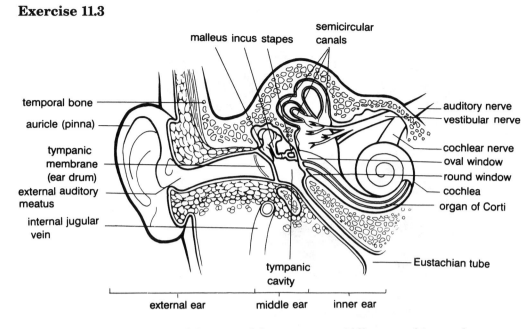

temporal bone

auricle (pinna)

tympanic membrane (ear drum)

external auditory meatus

internal jugular vein

malleus incus stapes

semicircular canals

auditory nerve

vestibular nerve

cochlear nerve

oval window

round window

cochlea

organ of Corti

Eustachian tube

tympanic cavity

external ear   middle ear   inner ear

Figure 11.3 Frontal diagram of the outer ear, middle ear, and internal ear. A section of the cochlear duct has been cut away to show the position of the organ of Corti.

Test Items

A. 1.d, 2.c, 3.c, 4.a, 5.c, 6.c, 7.c, 8.d, 9.d, 10.c, 11.c, 12.b, 13.c, 14.d, 15.d, 16.b, 17.d, 18.b, 19.a, 20.b.

B. 1.i, 2.e, 3.h, 4.b, 5.j, 6.d, 7.c, 8.f, 9.g, 10.a.
   1.h, 2.a, 3.c, 4.b, 5.f, 6.d, 7.e, 8.g, 9.i, 10.j.

C. 1.T, 2.T, 3.F, 4.F, 5.T, 6.F, 7.T, 8.F, 9.T, 10.F, 11.T, 12.F, 13.F, 14.T, 15.F, 16.T, 17.T, 18.T, 19.F, 20.T.

Special Senses

Across

1 inner nerve layer of the eye

3 disease of the outer eye resulting in increased intraocular pressure

5 the ear lobe

6 central focus

8 colored muscular part of the outer eye

9 middle vascular layer of the eye

10 network of nerves in the cochlea

13 farsightedness

14 photoreceptor chemical

15 receptors sensitive to chemical stimuli

19 a middle ear bone

Down

2 the eardrum

4 the outer connective tissue layer of the eye

5 pertaining to hearing

7 pertaining to vision

9 loss of transparency of the lens

11 curved inward

12 double vision

16 nearsightedness

17 curved outward

18 opening of the iris

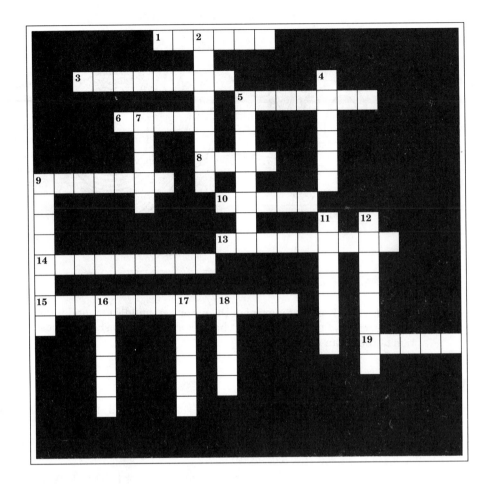

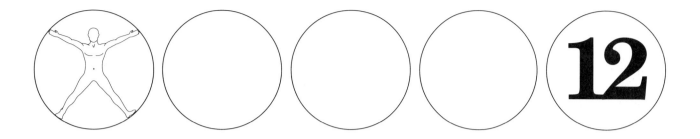

The Blood

I. CHAPTER SYNOPSIS

The central theme of this chapter is that blood maintains the constancy of the internal environment and, in the process, it too remains constant within limits—blood protein content, pH, blood pressure, and osmotic pressure are constant, for example.

Blood maintains the constancy of the tissues through the exchange of molecules that takes place in the capillaries. This exchange is an important portion of the chapter, and it is considered in detail.

The major concern of this chapter is to analyze the structure and function of blood and to note interrelations in maintaining homeostasis. This is developed through a study of the origin and functions of the formed elements in blood and a comparison of the location and composition of plasma and interstitial fluid. The transportation of respiratory gases, the reticulocyte count, the differential count, phagocytosis, the antigen-antibody response, blood clotting, clotting tests, and the ABO and Rh blood grouping systems are considered. Among the blood disorders discussed are several kinds of anemia, polycythemia, infectious mononucleosis, and leukemia.

The human body is continually subjected to stress by invasion of microorganisms (i.e., viruses and bacteria) whose growth, if unchecked, will disrupt the homeostasis and normal functioning of the body through the release of toxic materials and the destruction of cells. The human body has a number of passive (anatomical barriers such as the skin) and active (the immune system and phagocytic activity of the WBCs) defenses against these invaders. The body cells also wear out and must be removed and abnormal or mutant cell types that arise must be destroyed for they put stress on normal homeostasis. Thus, this chapter deals with the immune system and the physiological mechanisms that allow the body to recognize foreign material and stressful situations and to neutralize or eliminate them.

II. OBJECTIVES

After reading the chapter, the student should be able to:

- List the functions of blood.
- Describe the characteristics of blood.
- Define hematocrit.
- Explain the function of hemoglobin.
- Describe how an erythrocyte is produced and how it is destroyed.
- Differentiate between the intrinsic and extrinsic factors.
- Explain the relationship of hemolysis and a high bilirubin level.
- Distinguish among the kinds of leukocytes and their functions.
- Differentiate between leukopenia and leukocytosis.
- Define diapedesis and relate its role in inflammation.
- Describe a thrombocyte and explain its role in the clotting mechanism.
- Define immunity.
- Explain the role of B and T lymphocytes in the immune reaction.

III. IMPORTANT TERMS

Using your textbook, define the following terms:

agglutination (ah-gloot-in-ay'-shun) _____

_____

allergen (al'-ur-jen) _____

_____

anemia (ah-nee'-mee-ah) _____

_____

antibody (ant'-eh-bod-ee) _____

_____

anticoagulant (ant-eh-ko-ag'-yah-lunt) _____

_____

antigen (ant'-eh-jin) _____

_____

bilirubin (bil-eh-roo'-bin) _____

_____

coagulation (ko-ag-yah-lay'-shun) _____

_____

diapedesis (di-ah-pah-dee'-sis) _____

_____

erythrocyte (eh-rith'-rah-sight) _____

_____

extrinsic (ek-strin'-zik) _____

_____

hematocrit (hee-mat'-ah-krit) _____

_____

hemoglobin (hee'-mah-glo-bin) _____

_____

hemopoietic (hee-mah-poy-et'-ik) _____

_____

heparin (hep'-ah-rin) _____

_____

histamine (his'-tah-meen) _____

intrinsic (in-trin'-zik) _____

_____

leukocyte (lew'-kah-sight) _____

lysis (li'-sis) _____

platelet (plait'-lit) _____

_____

thrombocyte (throm'-bah-sight) _____

_____

thrombus (throm'-bus) _____

_____

IV. EXERCISES

Complete the following exercises in the order given. A precise set of terms
and diagrams has been chosen to describe the blood.

Exercise 12.1

Completion.   Write the name of the characteristic of blood in the space
provided.

Key:

arteries

bright red

dark red

density

flows

less

oxygenated

pH

6 L

slowly

temperature

thicker

veins

viscosity

volume

volume

water

weight

CHARACTERISTICS OF BLOOD

Blood is _____ (indicating that it is well _____)
in the _____ and _____ (_____
oxygenated) in the _____. It _____ four to
five times more _____ than water because it is four
to five times _____ (a property called _____).
Its specific gravity (_____ compared with water) varies between 1.045 to 1.065 (_____ is 1.000), and its
_____ and _____ values are 38°C (100.4°F)
and about 7.38, respectively. The _____ of blood in
the body has been measured in many ways but can be expected
to vary with the size of the body. A useful estimate is that its
_____ is 8 percent of the body's weight. The blood
_____ of a man of average size is approximately
_____ .

Key:
anticoagulants
cell
clotting
corpuscles
erythrocytes
55
formed
hematocrit
homogeneous
leukocytes
lower
opaque
percentage
plasma
plasma
thrombocytes
transparent
upper
venipuncture
yellow

Seen with the naked eye, blood appears to be _____ and _____ ; but on microscopic examination, it consists of _____ fragments and an intercellular fluid, the _____ .

When blood is obtained from a person's vein (by _____) and transferred to a test tube, and _____ is prevented by adding certain chemicals called _____ , it separates into two distinct layers. The _____ layer, which is _____ and _____ , contains most of the chemical components of the blood in solution; this is the _____ . Plasma constitutes about _____ percent of the blood's volume. The _____ portion of the blood sample consists of the _____ elements: the red blood cells (_____), the white blood cells (_____), and the platelets (_____). These are often called _____, meaning little bodies. The _____ of total blood volume contributed by these formed elements is called the _____ .

Exercise 12.2

Labeling. Write the name of the formed element in the space provided. Color each cell differently.

Key:
basophil
eosinophil
erythrocytes
lymphocyte
monocyte
neutrophil
thrombocytes

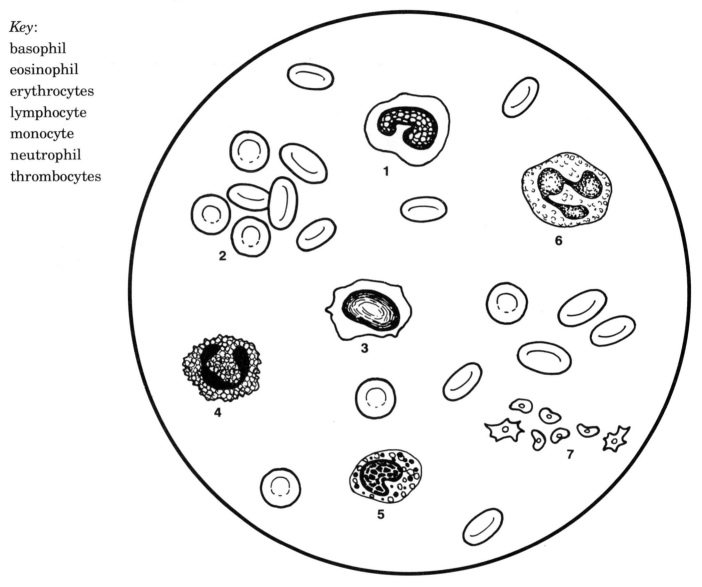

Figure 12.2 Microscopic blood smear.

1. _____   5. _____

2. _____   6. _____

3. _____   7. _____

4. _____

Exercise 12.3

Labeling. Write the name of the blood factor in the space provided. Color each area differently.

Key:

circulation

erythrocyte

excreted

extrinsic

gastric

hematopoietic

intrinsic

RBC

stimulate

WBC

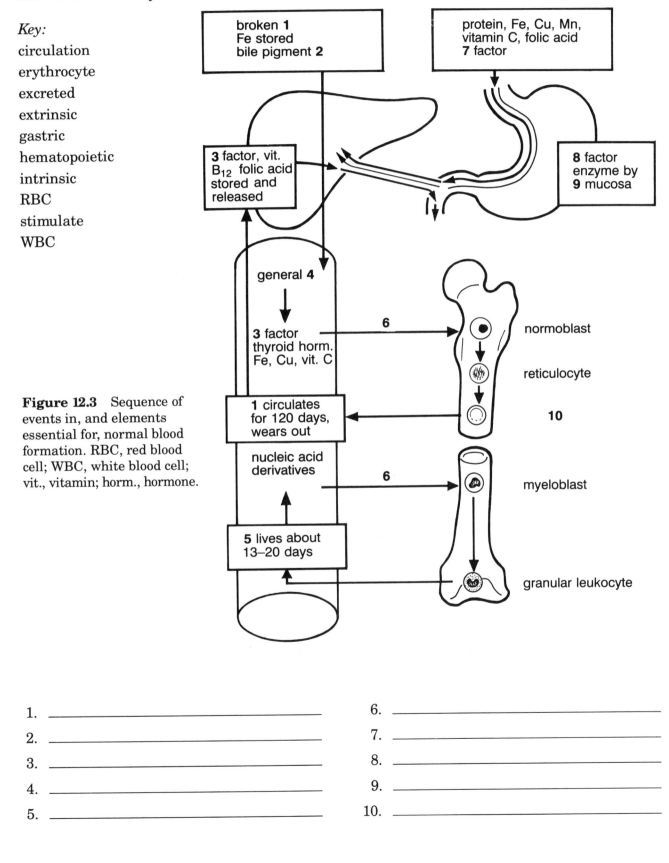

Figure 12.3   Sequence of events in, and elements essential for, normal blood formation. RBC, red blood cell; WBC, white blood cell; vit., vitamin; horm., hormone.

1. _____   6. _____

2. _____   7. _____

3. _____   8. _____

4. _____   9. _____

5. _____   10. _____

V. TEST ITEMS

A. *Multiple Choice.* There is only one answer that is either correct or most appropriate. Circle the answer that corresponds to the question.

1. Hemoglobin is
   a. carried in red blood cells.
   b. an oxygen transporter.
   c. made of protein and heme.
   d. all of the above

2. All of the following may be found in blood except
   a. fibrinogen.
   b. glucose.
   c. urea.
   d. glycogen.

3. The two enzymes in the clotting processes are
   a. fibrin and thromboplastin.
   b. thromboplastin and thrombin.
   c. platelets and fibrin.
   d. prothrombin and calcium.

4. A person with normal blood volume and white cell count but low red cell count or low hemoglobin content is suffering from
   a. hemophilia.
   b. anemia.
   c. leukemia.
   d. mononucleosis.

5. When a person receives blood of the wrong type,
   a. the white cells clot.
   b. hemorrhaging occurs.
   c. fibrinogen is activated and fibrin threads appear.
   d. agglutination occurs.

6. In a severe allergic reaction, anaphylactic shock is caused by the prolonged effects of
   a. histamine.
   b. epinephrine.
   c. heparin.
   d. fibrinogen.

7. The blood cell that is said to be immunologically competent, that is, capable of antibody formation, is the
   a. small lymphocyte.
   b. blood platelet.
   c. neutrophil.
   d. eosinophil.
   e. mast cell.

8. The most actively phagocytic cells and the most amoeboid cells are the
   a. monocytes.
   b. neutrophils.
   c. small lymphocytes.
   d. basophils.
   e. eosinophils.

9. When phagocytic cells are attracted toward an injured area, this is said to be an example of
   a. diapedesis.
   b. phototropism.
   c. the effect of gravity.
   d. positive chemotaxis.
   e. leukopenia.

10. The blood group composing the smallest percentage in the general white population is blood type
   a. A.
   b. B.
   c. AB.
   d. O.
   e. Rh–.

11. Blood type AB is the universal recipient because
   a. it has no antibodies.
   b. it has no antigens.
   c. it has all antigens.
   d. it has all antibodies.

12. An antibody is
   a. a compound that reacts with an antigen.
   b. a white corpuscle that phagocytizes invading bacteria.
   c. a carbohydrate.
   d. a platelet.

13. Immunity is
   a. the opposite to allergic.
   b. dependent on the proper functioning of the nervous system.
   c. dependent on the presence of antibodies.
   d. a factor that increases with age.

14. A vaccine contains
   a. penicillin.
   b. horse serum.
   c. treated antigens.
   d. antibodies.

15. Which type of white blood cell is most important for immunity?
   a. polymorphonuclear ones
   b. monocytes
   c. agranulocytes
   d. lymphocytes

16. Which of the following is most specifically responsible for antibody-mediated immunity?
   a. T cell
   b. B cell
   c. platelets
   d. all of the above

17. The most important function of T cells is
   a. phagocytosis.
   b. antibody production.
   c. rejecting tissue implants.
   d. forming B cells.

18. Blood serum
   a. clots.
   b. is blood plasma minus fibrinogen.
   c. is blood plasma from oxalated or citrated blood.
   d. is blood plasma minus all of the clotting elements.

19. Thromboplastin
   a. is found in blood platelets.
   b. converts fibrinogen into fibrin.
   c. converts prothrombin to thrombin in the presence of sodium ions.
   d. forms the network of the clot in which blood cells are trapped.

20. In type A blood, which antibody would normally be present in the plasma?
   a. a
   b. b
   c. and and b
   d. neither a nor b

B. *Matching Questions.*   Each of the phrases in COLUMN B refers to a word or phrase in COLUMN A. Insert the letter of the word or phrase from COLUMN B that best describes it. Some words or phrases may be used more than once or not at all.

	Column A		*Column B*
1. ___	ADH	**a.**	derived from B lymphocytes
2. ___	Rh incompatibility	**b.**	release histamine
3. ___	plasma cells	**c.**	phagocytize bacteria
4. ___	transfusion reaction	**d.**	a hormone released during stress
5. ___	cortisol	**e.**	stimulates gluconeogenesis
6. ___	destroy cancer cells	**f.**	the fight-or-flight response
7. ___	autoimmune	**g.**	antibodies against "self" tissue
8. ___	active immunity	**h.**	atopic allergy
9. ___	hay fever	**i.**	erythroblastosis fetalis
10. ___	basophils	**j.**	a special type of tissue rejection
11. ___	sympathetic stimulation	**k.**	"self" protein molecules that are antigenic
12. ___	passive immunity	**l.**	cell-mediated immunity
13. ___	clones	**m.**	recipient receives performed antibodies
14. ___	histocompatibility antigens	**n.**	different populations of B cells
15. ___	macrophages	**o.**	antibodies are built up as a result of actual contact

Column A		*Column B*
1. ___ agglutinogens	**a.**	antibody
2. ___ fibrinogen	**b.**	Vitamin K
3. ___ platelets	**c.**	blood clot
4. ___ red bone marrow	**d.**	anticoagulant
5. ___ erythropoietin	**e.**	differentiation of hemocytoblast
6. ___ spleen	**f.**	development of white blood cells
7. ___ heparin	**g.**	manufactures RBCs during fetal life
8. ___ agglutinin	**h.**	manufactures RBCs in adults
9. ___ leukopoiesis	**i.**	thromboplastin
10. ___ prothrombin	**j.**	antigens

C. *True-False.* Place a *T* or *F* in the space provided.

___ 1. The release of histamine from damaged tissues produces inflammation (reddening) by increasing the permeability of capillaries to red blood cells that accumulate in the extracellular spaces surrounding the damaged tissue.

___ 2. When a blood vessel is severed or injured, its immediate response is to constrict.

___ 3. The event transforming blood into a solid clot is the conversion of plasma fibrinogen to fibrin.

___ 4. Vitamin K is an essential cofactor in the liver's synthesis of prothrombin and the plasma factors.

___ 5. Bilirubin is a breakdown component of hemoglobin.

___ 6. About 70 percent of the total body iron is in hemoglobin.

___ 7. Vitamin B regulates the rate of erythrocyte production.

___ 8. Polymorphonuclear granulocytes refers to the three types of blood cells with lobulated nuclei and cytoplasmic granules.

___ 9. The cells of the body believed to form antibodies are called neutrophils.

___ 10. Any substance capable of stimulating antibody production is an antigen.

___ 11. Substances such as heparin and dicumarol that promote fibrinolysis are classified as anticoagulants.

___ 12. Blood type O is referred to as the universal recipient because it has neither A nor B antibodies in its plasma.

___ 13. The normal range for clotting time is between 5 and 15 minutes.

___ 14. In contrast with plasma, both lymph and interstitial fluid lack thrombocytes and erythrocytes.

_____ **15.** Excessive loss of red blood cells through bleeding may result in sickle cell anemia.

_____ **16.** An abnormal increase in the number of red blood cells is referred to as hemolysis.

_____ **17.** Slight fever, sore throat, stiff neck, cough, malaise, and a high monocyte and leucocyte count are indicative of a disorder of white blood cells called leukemia.

_____ **18.** The antibodies in blood plasma are referred to as anticoagulants.

_____ **19.** Antibodies are all composed of polypeptide chains.

_____ **20.** The lymphocyte is the largest of the white blood cells.

Answer Sheet—Chapter 12

Exercise 12.1

CHARACTERISTICS OF BLOOD

Blood is <u>bright red</u> (indicating that it is well <u>oxygenated</u>) in the <u>arteries</u> and <u>dark red</u> (<u>less</u> oxygenated) in the <u>veins</u>. It flows four to five times more <u>slowly</u> than water because it is four to five times <u>thicker</u> (a property called <u>viscosity</u>). Its specific gravity (<u>density</u> compared with water) varies between 1.045 to 1.065 (<u>water</u> is 1.000), and its <u>temperature</u> and <u>pH</u> values are 38°C (100.4°F) and about 7.38, respectively. The <u>volume</u> of blood in the body has been measured in many ways but can be expected to vary with the size of the body. A useful estimate is that its <u>weight</u> is 8 percent of the body's weight. The blood <u>volume</u> of a man of average size is approximately <u>6 L</u>.

Seen with the naked eye, blood appears to be <u>opaque</u> and <u>homogeneous</u>; but on microscopic examination, it consists of <u>cell</u> fragments and an intercellular fluid, the <u>plasma</u>.

When blood is obtained from a person's vein (by <u>venipuncture</u>) and transferred to a test tube, and <u>clotting</u> is prevented by adding certain chemicals called <u>anticoagulants</u>, it separates into two distinct layers. The <u>upper</u> layer, which is <u>yellow</u> and <u>transparent</u>, contains most of the chemical components of the blood in solution; this is the <u>plasma</u>. Plasma constitutes about <u>55</u> percent of the blood's volume. The <u>lower</u> portion of the blood sample consists of the <u>formed</u> elements: the red blood cells (<u>erythrocytes</u>), the white blood cells (<u>leukocytes</u>), and the platelets (<u>thrombocytes</u>). These are often called <u>corpuscles</u>, meaning little bodies. The <u>per-centage</u> of total blood volume contributed by these formed elements is called the <u>hematocrit</u>.

Exercise 12.2

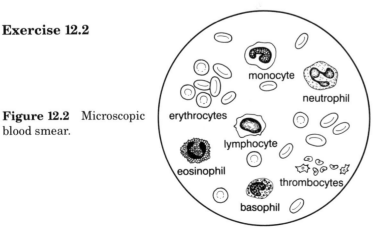

Figure 12.2 Microscopic blood smear.

Exercise 12.3

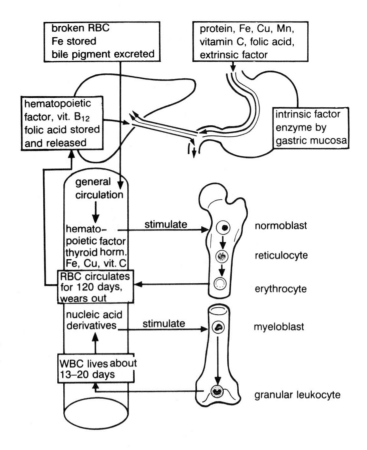

Figure 12.3 Sequence of events in, and elements essential for, normal blood formation. RBC, red blood cell; WBC, white blood cell; vit., vitamin; horm., hormone.

Test Items

A. 1.d, 2.d, 3.b, 4.b, 5.d, 6.a, 7.a, 8.b, 9.d, 10.c, 11.a, 12.a, 13.c, 14.c, 15.d, 16.b, 17.c, 18.b, 19.a, 20.b.

B. 1.d, 2.i, 3.a, 4.j, 5.e, 6.l, 7.g, 8.o, 9.h, 10.b, 11.f, 12.m, 13.n, 14.k, 15.c.
1.j, 2.c, 3.i, 4.h, 5.e, 6.g, 7.d, 8.a, 9.f, 10.b.

C. 1.F, 2.T, 3.T, 4.T, 5.T, 6.T, 7.F, 8.T, 9.F, 10.T, 11.T, 12.F, 13.T, 14.T, 15.F, 16.F, 17.F, 18.F, 19.T, 20.F.

Chapter 12

The Blood

Across

2 a substance formed in response to an antigen

6 a foreign substance creating sensitivity

8 —— to rupture

9 a substance of inside origin

11 formed element content of the blood

13 a substance that stimulates the creation of antibodies

14 prevents clotting of blood

15 a condition of low blood cell count

16 of outside origin

Down

1 an iron-protein molecule that carries gases

2 a substance that prevents blood clotting

3 a stationary blood clot

4 a waste produced from cell destruction

5 a red blood cell

10 passage of leucocytes through blood vessels

11 a vasodilator

12 clotting of blood

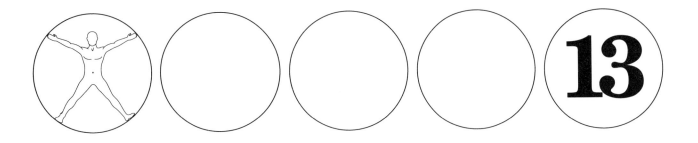

The Heart

I. CHAPTER SYNOPSIS

Chapter 13 presents the major anatomical and physiological features of the heart. Among the principal physiological aspects considered are the initiation and conduction of the heartbeat, the electrocardiogram, the heart's blood supply, the major events of the cardiac cycle, heart sounds, cardiac output and stroke volume, regulation of heart rate and circulatory shock, and homeostasis.

As pointed out in the previous chapter, in a multicellular organism such as a human being, diffusion from the surface of the body is too slow a process to deliver nutrients and oxygen to the cells of the body and to remove waste products and carbon dioxide. The circulatory system provides the body with a mechanism for rapidly exchanging matter between the external and internal environments. The heart, acting as a pump, provides the force necessary to produce a bulk flow of blood that is channeled to the tissues through blood vessels where the exchange of matter between the blood and interstitial fluid occurs.

II. OBJECTIVES

After reading the chapter, the student should be able to:

- Identify the structures of the heart and trace the cardiac cycle.

- Describe the two sets of heart valves and explain their function in cardiac blood flow.

- Explain why the sinoatrial node is the pacemaker of the heart.
- Define coronary artery disease.
- List and explain at least two kinds of heart attack.
- Describe the cardiac reflex in relation to neurochemical control.
- Explain congenital heart disease and give examples.

III. IMPORTANT TERMS

Using your textbook, define the following terms:

apex (ay'-peks) _____

_____

atrium (ay'-tree-um) _____

_____

bicuspid (bi-kus'-pid) _____

_____

bradycardia (braid-eh-kahrd'-ee-ah) _____

_____

circumflex (sur'-kum-fleks) _____

_____

coronary (kor'-ah-ner-ee) _____

_____

diastole (di-as'-tah-lee) _____

_____

endocardium (en-do-kahrd'-ee-um) _____

_____

epicardium (ep-ah-kard'-ee-um) _____

_____

foramen (fah-ray'-men) _____

infarction (in-fahrk'-shun) _____

_____

ischemia (is-kee'-mee-ah) _____

_____

murmur (mur'-mur) _____

_____

myocardium (mi-ah-kahrd'-ee-um) _____

_____

pericardium (per-ah-kahrd'-ee-um) _____

_____

Purkinje (pur-kin'-jee) _____

_____

septum (sep'-tum) _____

_____

sinoatrial (si-no-ay'-tree-ul) _____

_____

stenoses (stah-no'-sis) _____

_____

systole (sis'-tah-lee) _____

_____

tachycardia (tak-eh-kahrd'-ee-ah) _____

_____

tricuspid (tri-kus'-pid) _____

_____

umbilical (um-bil'-i-kul) _____

_____

valve (valv) _____

_____

ventricle (ven'-tri-kul) _____

_____

IV. EXERCISES

Complete the following exercises in the order given. A precise set of terms
and diagrams has been chosen to describe the heart.

Exercise 13.1

Labeling.   Write the name of the structure in the space provided. Color the
right and left sides of the heart blue and red, respectively.

Key:

aorta	atrium	pulmonary	superior	veins
aortic	bicuspid	semilunar	tricuspid	ventricle
apex	inferior	septum	valve	

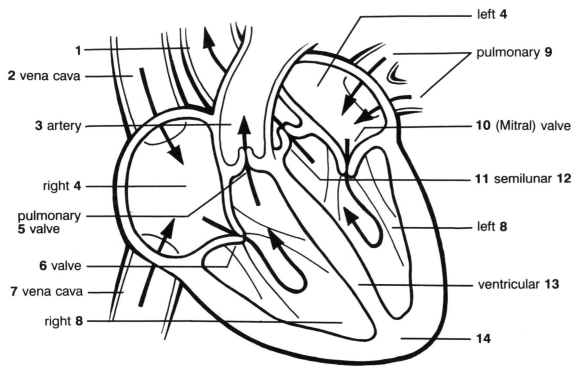

Figure 13.1   Diagram of the heart. Arrows indicate direction of blood flow.

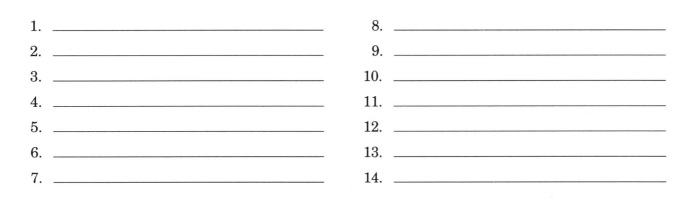

1. _____   8. _____

2. _____   9. _____

3. _____   10. _____

4. _____   11. _____

5. _____   12. _____

6. _____   13. _____

7. _____   14. _____

Exercise 13.2

Labeling.   Write the name of the structure in the space provided. Color the excitation system black and the heart tissue light red.

Key:

atrium	interventricular	sinoatrial
atrioventricular	myocardium	ventricle
His	Purkinje	

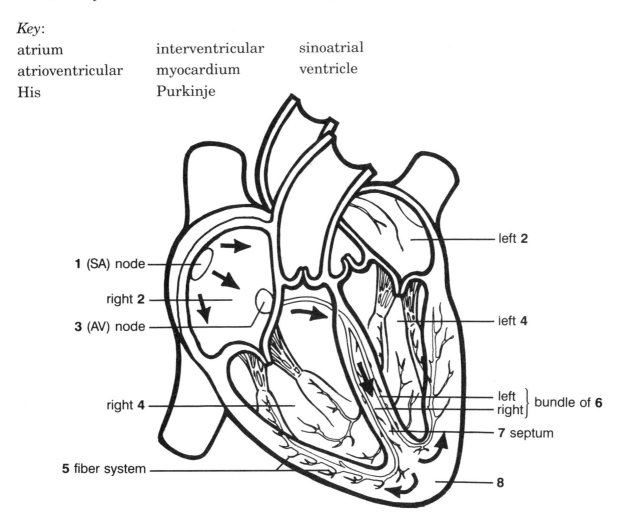

1 (SA) node

right **2**

3 (AV) node

right **4**

5 fiber system

left **2**

left **4**

left } bundle of **6**
right }

7 septum

8

Figure 13.2   Excitation-conduction system. Diagram showing how an impulse is conducted from its point of origin, the right atrium, to its destination, the myocardium.

1. _____

2. _____

3. _____

4. _____

5. _____

6. _____

7. _____

8. _____

Exercise 13.3

Labeling. Write the name of the structure in the space provided. Color the arteries, veins, pericardium, and fat, respectively.

Key:

aorta
aortic
artery
atrium
brachiocephalic
carotid
coronary
circumflex
pericardium
pulmonary
sinus
subclavian
superior
veins
vena cava
ventricle

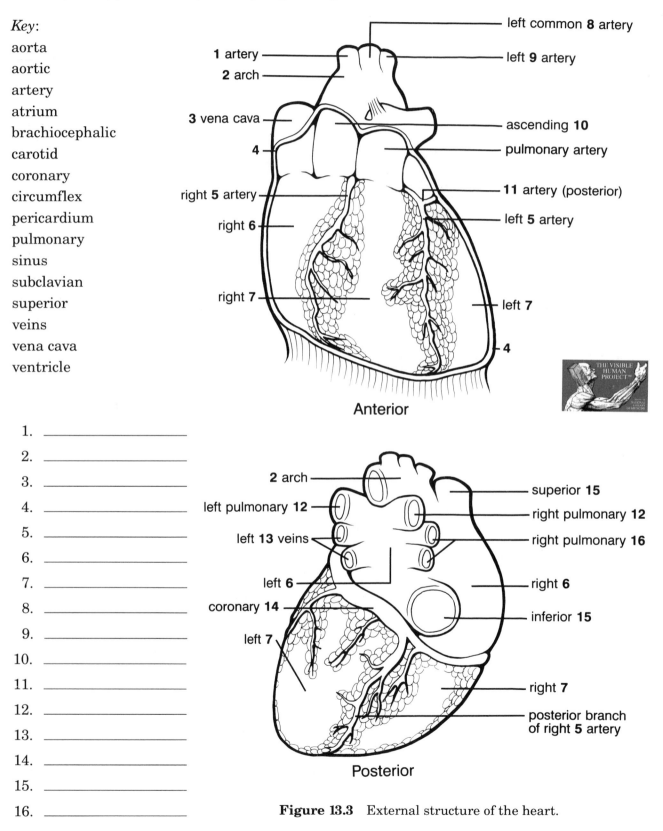

1. _____
2. _____
3. _____
4. _____
5. _____
6. _____
7. _____
8. _____
9. _____
10. _____
11. _____
12. _____
13. _____
14. _____
15. _____
16. _____

Figure 13.3   External structure of the heart.

Exercise 13.4

Labeling.   Write the name of the structure in the space provided. Color the different vessels and organs.

Key:

aorta
arch
arteriosus
atrium
foramen
iliac
liver
lung
placenta
pulmonary
umbilical
vena cava
venosus
ventricle

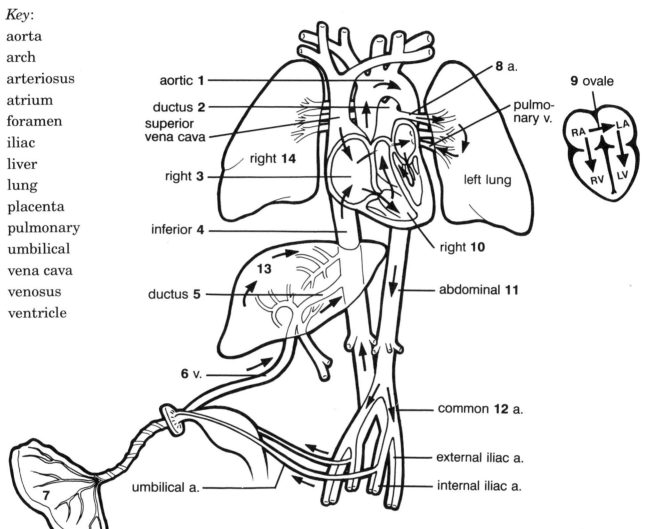

Figure 13.4   Plan of circulation in a mature fetus. Arrows indicate the direction of blood flow. The inset is a diagram of the blood flow in the fetal heart. Note the blood flow from the right atrium into both the right ventricle and left atrium through the foramen ovale. RA, right atrium; LA, left atrium; RV, right ventricle; LV, left ventricle.

1. _____
2. _____
3. _____
4. _____
5. _____
6. _____
7. _____

8. _____
9. _____
10. _____
11. _____
12. _____
13. _____
14. _____

V. TEST ITEMS

A. *Multiple Choice.*   There is only one answer that is either correct or most appropriate. Circle the answer that corresponds to the question.

1. Which of the following chambers pumps blood into the systemic circulatory system?
   a. right atrium
   b. left atrium
   c. right ventricle
   d. left ventricle

2. Freshly oxygenated blood is received by the
   a. right atrium.
   b. left atrium.
   c. right ventricle.
   d. left ventricle.

3. The membranous sac surrounding the heart is known as the
   a. pericardium.
   b. endocardium.
   c. myocardium.
   d. epicardium.

4. The P wave of the EKG correlates with the
   a. depolarization of the AV node.
   b. atrial depolarization.
   c. ventricular repolarization.
   d. depolarization of the ventricle.

5. Which of the following correctly depicts the functional differences between the left and right heart?
   a. The right heart works against greater resistance than the left heart.
   b. The left heart pumps more blood than the right heart.
   c. Starling's law of the heart applies to the left ventricle but not to the right ventricle.
   d. The partial pressure of oxygen in the blood of the left ventricle is lower than in the right ventricle.

6. Venous blood is received by the
   a. right atrium.
   b. left atrium.
   c. right ventricle.
   d. left ventricle.

7. The two distinct heart sounds, described phonetically as lubb and dupp, represent the
   a. contraction of the ventricles and the relaxation of the atria.
   b. contraction of the atria and the relaxation of the ventricles.
   c. closing of the atrioventricular and semilunar valves.
   d. surging of blood into the pulmonary artery and aorta.

8. Which of the following sequences best represents the course of a nerve impulse through the heart?
   1. sinoatrial node
   2. atrioventricular node
   3. Purkinje fibers
   4. bundle of His
   a. 1, 4, 3, 2
   b. 1, 2, 3, 4
   c. 1, 3, 2, 4
   d. 1, 2, 4, 3

9. The atrioventricular valve on the same side of the heart as the origin of the aorta is the
   a. aortic semilunar.
   b. tricuspid.
   c. mitral.
   d. pulmonary semilunar.

10. During atrial systole, all the following occur *except*
    a. deoxygenated blood passes into the right ventricle.
    b. oxygenated blood passes into the left ventricle.
    c. the ventricles are in diastole.
    d. the semilunar valves are open.

11. A major difference between atria and ventricles of the heart is that the former have
    a. a layer of myocardium.
    b. thinner walls.
    c. papillary muscles.
    d. a lining of endocardium.

12. An operation that attempts to repair an atrial septal defect would be carried out on the
    a. ductus arteriosus.
    b. ductus venosus.
    c. foramen ovale.
    d. interventricular septum.

13. A congenital heart disorder that results in cyanosis as a result of the mixing of oxygenated and deoxygenated blood is
    a. patent ductus arteriosus.
    b. atrial septal defect.
    c. ventricular septal defect.
    d. valvular stenosis.

14. A myocardial infarction results in
    a. the death of an area of the aorta.
    b. an accelerated rate of hemopoiesis.
    c. rapid cell division of the layers of the pericardium.
    d. death of an area of the heart muscle.

15. A "blue baby" is probably
    a. suffering from rheumatic heart disease.
    b. suffering from arterial sclerosis.
    c. not getting enough oxygenated blood throughout the body.
    d. not getting enough liver in the body.

16. A pulse rate of 100 times per minute indicates
    a. bradycardia.
    b. myocardia.
    c. tachycardia.
    d. endocardia.

17. The blood flows from the lungs into the heart's
    a. right atrium.
    b. left ventricle.
    c. atrioventricular valve.
    d. left atrium.

18. Systole occurs when
    a. the heart muscle contracts.
    b. the heart muscle relaxes.
    c. the atrioventricular valve closes.
    d. the semilunar valve closes.

19. The heart beat originates in the
    a. AV node.
    b. pacemaker.
    c. autonomic nervous system.
    d. pericardium.

20. What is the normal heart rate for a young adult?
    a. 100 beats per minute
    b. 90 beats per minute
    c. 70 beats per minute
    d. 50 beats per minute

B. *Matching Questions.*   Each of the phrases in COLUMN B refers to a word or phrase in COLUMN A. Insert the letter of the word or phrase from COLUMN B that best describes it. Some words or phrases may be used more than once or not at all.

	Column A		*Column B*
1. ___	norepinephrine	**a.**	increases cardiac output
2. ___	intrinsic control	**b.**	increases ventricular and atrial contractility
3. ___	SA node	**c.**	the external pressure surrounding the heart
4. ___	systole	**d.**	the direct proportion between the diastolic volume of the heart and the force of the contraction
5. ___	T wave		
6. ___	ectopic foci	**e.**	an inherent property of cardiac muscle
7. ___	epinephrine	**f.**	under constant influence of nerves and hormones
8. ___	intercalated disks	**g.**	a leaky valve
9. ___	murmur	**h.**	ventricular action
10. ___	fibrillation	**i.**	ventricular repolarization
11. ___	apex	**j.**	disorganized contractions
12. ___	intrathoracic	**k.**	caffeine may cause
13. ___	myocardium	**l.**	"tight junctions"
14. ___	Starling's law of the heart	**m.**	the tip of the left ventricle
15. ___	arteries	**n.**	heart muscle
		o.	carries blood away from the heart

Column A	*Column B*
1. ___ tachycardia	a. mechanical device for applying electrical shock to the heart
2. ___ defibrillator	
3. ___ cardiac arrest	b. heart sound produced by blood passing through a valve or opening caused by a septal defect
4. ___ cyanosis	c. rapid heart rate
5. ___ bradycardia	d. slightly bluish skin coloration due to oxygen deficiency in systemic blood
6. ___ murmur	e. complete stoppage of the heartbeat
	f. slow heartbeat

C. *True-False.*   Place a *T* or *F* in the space provided.

___ 1. The reflex that controls venous blood pressure is the carotid sinus reflex.

___ 2. Sympathetic stimulation of the heart brings about an increase in heart rate.

___ 3. The phenomenon by which the length of the cardiac muscle fiber determines the force of contraction is called Starling's law of the heart.

___ 4. In an ECG, the QRS wave represents the spread of an electrical impulse through the atria.

___ 5. The mass of conducting cells located in the right atrium that serves as the pacemaker is called the sinoatrial node.

___ 6. During ventricular systole, the ventricles are in a period of relaxation.

___ 7. Under normal circumstances, the right side of the heart contains only deoxygenated blood.

___ 8. The condition known as mitral stenosis (narrowing) hinders blood's flow from the right atrium into the right ventricle.

___ 9. The coronary arteries are the first branches off the thoracic aorta.

___ 10. Although anastomoses between coronary arteries are numerous, they are small, hence the many deaths from coronary obstruction.

___ 11. To catheterize the right heart the catheter must be introduced through a vein.

___ 12. Whereas the sinoatrial node serves as the natural pacemaker of the heart, the atrioventricular node coordinates ventricular contractions with atrial contractions.

___ 13. The left ventricle has to pump more blood than the right ventricle because it has to pump the blood through the whole body, not just through the pulmonary system.

___ **14.** The cardiovascular integrating center is located in the medulla of the brain stem.

___ **15.** The force of contraction of the left ventricle is greater than the force of contraction of the right ventricle.

Answer Sheet—Chapter 13

Exercise 13.1

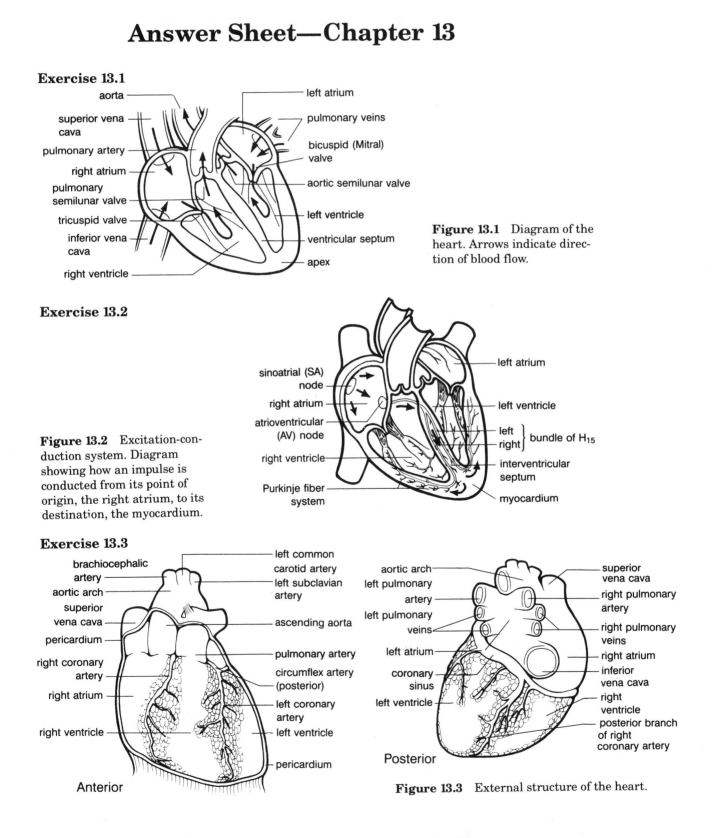

Figure 13.1 Diagram of the heart. Arrows indicate direction of blood flow.

Exercise 13.2

Figure 13.2 Excitation-conduction system. Diagram showing how an impulse is conducted from its point of origin, the right atrium, to its destination, the myocardium.

Exercise 13.3

Figure 13.3 External structure of the heart.

Exercise 13.4

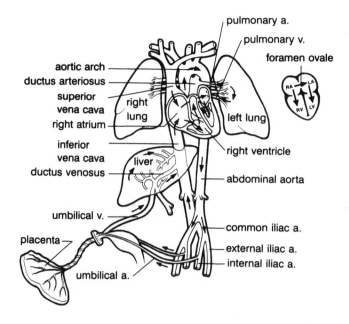

aortic arch
ductus arteriosus
superior vena cava
right atrium
inferior vena cava
ductus venosus
umbilical v.
placenta
umbilical a.
right lung
liver
pulmonary a.
pulmonary v.
foramen ovale
left lung
right ventricle
abdominal aorta
common iliac a.
external iliac a.
internal iliac a.
RA LA
RV LV

Figure 13.4 Plan of circulation in a mature fetus. Arrows indicate the direction of blood flow. The inset is a diagram of the blood flow in the fetal heart. Note the blood flow from the right atrium into both the right ventricle and left atrium through the foramen ovale. RA, right atrium; LA, left atrium; RV, right ventricle; LV, left ventricle.

Test Items

A. 1.d, 2.b, 3.a, 4.b, 5.a, 6.a, 7.c, 8.d, 9.c, 10.d, 11.b, 12.c, 13.b, 14.d, 15.c, 16.c, 17.d, 18.a, 19.b, 20.c.

B. 1.b, 2.e, 3.f, 4.h, 5.i, 6.k, 7.a, 8.l, 9.g, 10.j, 11.m, 12.c, 13.n, 14.d, 15.o.
1.c, 2.a, 3.e, 4.d, 5.f, 6.b.

C. 1.F, 2.T, 3.T, 4.F, 5.T, 6.F, 7.T, 8.F, 9.F, 10.F, 11.T, 12.T, 13.F, 14.T, 15.T.

The Heart

Across

1 narrowing of an opening

3 microscopic nerve fibers in the myocardium

4 an opening

5 connective tissue covering-the 'sac'

7 a double-lipped heart valve

10 pertaining to the central region of the abdomen

17 to rest

20 a branch of the left coronary artery

21 receiving chamber of the heart

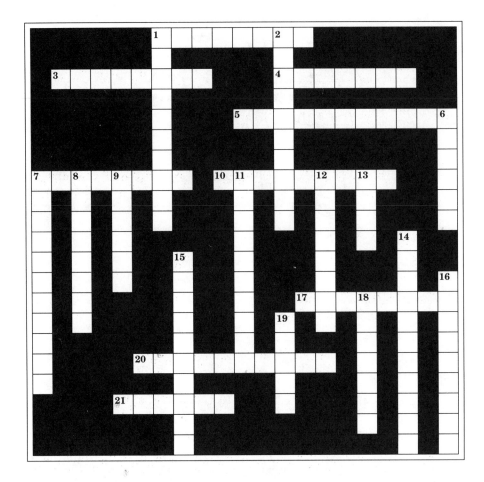

Down

1 specialized neuro-muscular tissue of the heart

2 tissue death due to loss of blood

6 abnormal heart sound due to a leaky valve

7 a slower than normal heart rate

8 blood supply to the heart

9 a connective tissue wall

11 heart muscle

12 lack of blood in the area of muscle

13 inferior tip of the heart

14 connective tissue inner lining of the heart

15 connective tissue covering of the myocardium

16 the pumping chamber of the heart

18 action, depolarization

19 a multi-lipped opening between two chambers

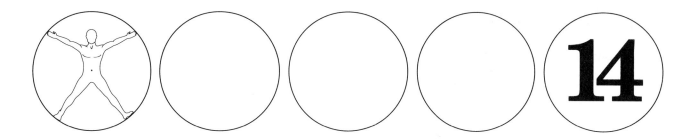

Circulation

I. CHAPTER SYNOPSIS

In the previous chapter, the control of the heart was emphasized. After blood leaves the heart, a variety of control systems and pathways regulate the distribution of blood to the tissues in proportion to the varying metabolic demands of the organs and systems of the body. These control mechanisms are integrated into the overall design of the circulatory system. The overall design, outside of the heart, consists of the arteries, arterioles, capillaries, veins, and the integrated lymphatic system. Thus, this chapter concerns itself with the basic anatomic and physiologic modalities of the vascular system in maintaining overall circulatory homeostasis.

II. OBJECTIVES

After reading the chapter, the student should be able to:

- Identify the structural differences between an artery, vein, and capillary.
- Identify the major arteries and veins of the systemic circulation.
- Identify the major pressure and pulse points of the body.
- Differentiate between pulmonary and systemic circulation.
- Describe the anatomical differences between the hepatic portal and fetal circulatory systems.

III. IMPORTANT TERMS

Using your textbook, define the following terms:

alveolar (al-vee'-ah-lahr) _____

_____

aneurysm (an'-yah-riz-em) _____

_____

apoplexy (ap'-ah-plek-see) _____

_____

arterioles (ahr-teer'-ee-ols) _____

_____

arteriosclerosis (ahr-teer-ee-o-sklah-ro'-sis) _____

_____

artery (ahrt'-ah-ree) _____

_____

brachial (bray'-kee-ul) _____

_____

capillary (kap'-ah-ler-ee) _____

_____

cerebrovascular (sah-ree-bro-vas'-kew-lur) _____

_____

hemorrhage (hem'-ah-rij) _____

_____

hepatic (heh-pat'-ik) _____

_____

hypertension (hi-pur-ten'-shun) _____

_____

oncotic (ong-kot'-ik) _____

_____

oxygenated (ock'-seh-jah-nate-ed) _____

_____

portal (port'-ul) _____

_____

pulmonary (pul'-mah-ner-ee) _____

_____

pulse (puls) _____

_____

saphenous (sah-fee'-nus) _____

_____

shock (shok) _____

_____

sphygmomanometer (sfig-mo-mah-nom'-et-ur) _____

_____

stethoscope (steth'-ah-skope) _____

_____

varicose (var'-ah-kose) _____

_____

vascular (vas'-kew-lur) _____

_____

vein (vain) _____

_____

venule (ven'-yool) _____

_____

vessel (ves'-ul) _____

_____

IV. EXERCISES

Complete the following exercises in the order given. A precise set of terms and diagrams has been chosen to describe the circulatory system.

Exercise 14.1

Labeling. Write the name of the artery in the space provided. Color the arteries red.

Key:
aortic
axillary
brachial
carotid
femoral
iliac
mesenteric
pedis
peroneal
popliteal
radial
renal
subclavian
temporal
tibial
ulnar

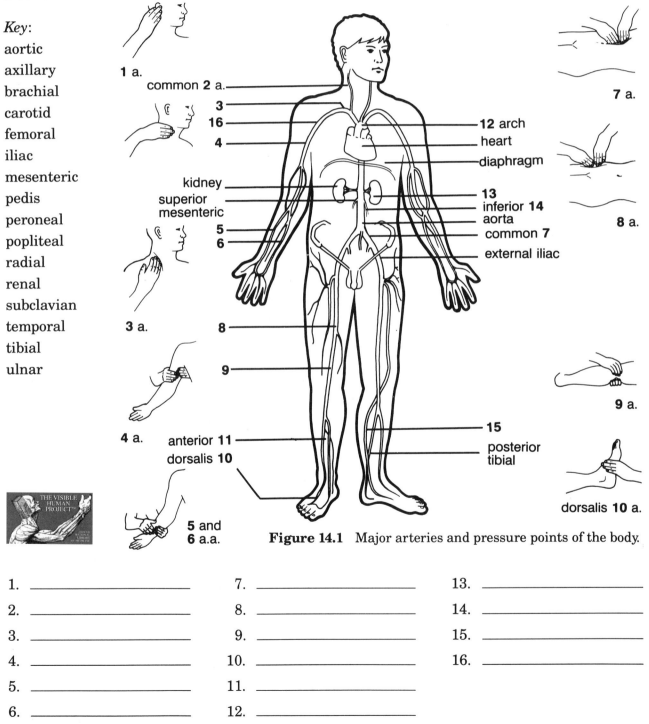

Figure 14.1 Major arteries and pressure points of the body.

1. _____   7. _____   13. _____

2. _____   8. _____   14. _____

3. _____   9. _____   15. _____

4. _____   10. _____   16. _____

5. _____   11. _____

6. _____   12. _____

Exercise 14.2

Labeling. Write the name of the vein in the space provided. Color the veins blue.

Key:
axillary
basilic
brachial
femoral
iliac
innominate
jugular
popliteal
radial
renal
saphenous
subclavian
tibial
ulnar
vena cava

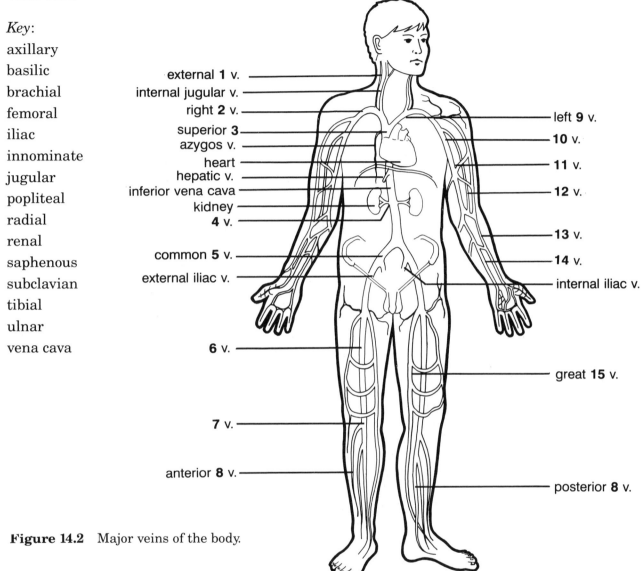

Figure 14.2   Major veins of the body.

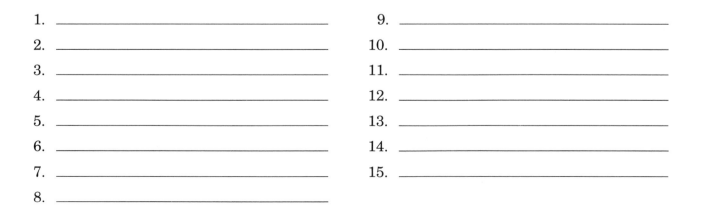

1. _____
2. _____
3. _____
4. _____
5. _____
6. _____
7. _____
8. _____

9. _____
10. _____
11. _____
12. _____
13. _____
14. _____
15. _____

Exercise 14.3

Labeling.   Write the name of the vein in the space provided. Color the veins blue and the organs various colors.

Key:
hepatic
iliac
mesenteric
portal
splenic
vena cava

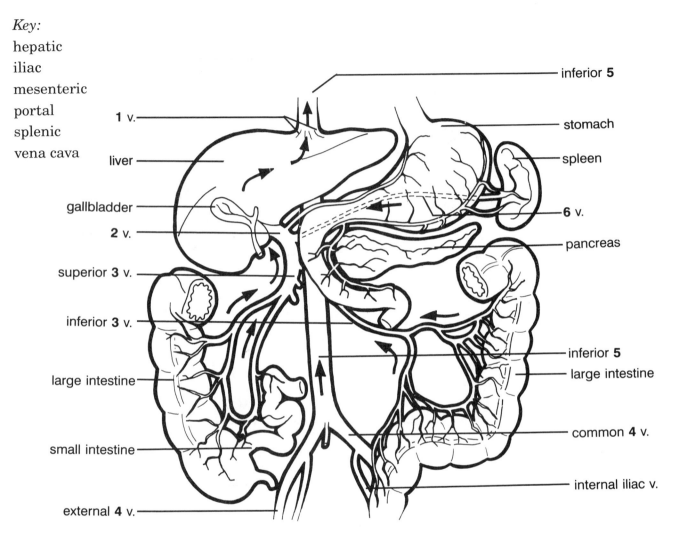

Figure 14.3   The hepatic portal system. Arrows indicate the direction of flow in all abdominal organs, through the liver, to the superior vena cava.

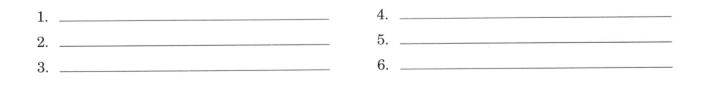

1. _____

2. _____

3. _____

4. _____

5. _____

6. _____

Exercise 14.4

Completion. Write the missing term in the space provided.

Key:

alveolar	heart
arterioles	leaves
artery	left
artery	lungs
blood	lungs
branches	pulmonary
bronchi	returns
capillary	right
divides	right
follows	veins

PULMONARY CIRCULATION

The _____ circulation carries _____ from the _____ to the _____ and back to the heart again. The blood _____ the heart by way of the large pulmonary _____ and _____ through the pulmonary _____ .

Originating in the _____ ventricle the pulmonary _____ passes upward and backward for a short distance and then _____ into _____ and _____ branches that enter the _____ . Within each lung the artery _____ widely, and each branch closely _____ the subdivisions of the _____ . The terminal _____ open into the _____ networks that approximate the _____ air sacs.

Key:

artery
artery
away
carbon dioxide
carbon dioxide
carry
capillary
condition
direction
left
lung
named
oxygenated
oxygenated
returning
to
two
vein
veins
venules
vessels

The pulmonary _____ entering the _____ atrium have _____ branches from each _____ . Their smallest tributaries arise as _____ from the _____ networks of the alveolar air sacs.

Of paramount importance here is the _____ of the blood in these vessels. As we have seen, blood _____ are _____ according to the _____ of flow to or from the heart—not in terms of what they _____ . Generally, a vessel leaving the heart is called an _____ and carries _____ blood (red); a vessel _____ to the heart is called a vein and carries _____-laden blood (blue). However, in the pulmonary circulation only, the pulmonary _____ carries _____-laden blood _____ from the heart (to the lungs), and the pulmonary _____ carries _____ blood _____ the heart (from the lungs).

V. TEST ITEMS

A. *Multiple Choice.* There is only one answer that is either correct or most appropriate. Circle the answer that corresponds to the question.

1. A rate of blood flow of 0.5 mm per sec would indicate that the blood is flowing through a (an)
   a. large artery.
   b. vein.
   c. arteriole.
   d. capillary.

2. The opening between the two atria in the fetus is called the
   a. foramen ovale.
   b. ductus venosus.
   c. ductus arteriosus.
   d. umbilical artery.

3. In taking blood pressure, the artery most commonly used is the
   a. radial.
   b. brachial.
   c. femoral.
   d. carotid.

4. Arterial pressure increases if
   a. peripheral resistance fails.
   b. peripheral resistance increases.
   c. viscosity decreases.
   d. vasodilation results.

5. The arterial circle of Willis is
   a. the circle around the stomach made by the celiac and gastric arteries.
   b. the circle formed by the arch of the aorta.
   c. a reference to the return of blood through the portal system.
   d. at the base of the brain.

6. The longest vein in the body, a superficial vein of the leg and thigh, is subject to enlargement called varicose veins. It is the
   a. femoral vein.
   b. popliteal vein.
   c. saphenous vein.
   d. tibial vein.

7. The principal effect of reducing the elasticity of the large arteries is to
   a. increase systolic and decrease diastolic pressure.
   b. decrease systolic pressure.
   c. increase diastolic pressure.
   d. do none of the above.

8. A reduction in oxygen concentration in the blood brings about vasoconstriction by acting principally on
   a. chemoreceptors in the aorta and carotid arteries.
   b. chemoreceptors in the medulla.
   c. arterial baroreceptors.
   d. none of the above

9. Reduction in blood volume often produces
   a. hypertension.
   b. shock.
   c. vasodilation.
   d. purpura.

10. Which statement best describes arteries?
    a. All carry oxygenated blood to the heart.
    b. All contain valves to prevent the backflow of blood.
    c. All carry blood away from the heart.
    d. Only large arteries are lined with endothelium.

11. Which statement is *not* true of veins?
    a. They have less elastic tissue and smooth muscle than arteries.
    b. They contain more fibrous tissue than arteries.
    c. Most veins in the extremities have valves.
    d. They always carry deoxygenated blood.

12. All arteries of the systemic circulation branch from the
    a. aorta.
    b. pulmonary artery.
    c. superior vena cava.
    d. circle of Willis.

13. A thrombus in the first branch of the aortic arch would affect the flow of blood to the
    a. left side of the head and neck.
    b. myocardium of the heart.
    c. right side of the head and neck and right upper extremity.
    d. left upper extremity.

14. A stroke occurs when blood cannot get to the
    a. heart muscles.
    b. kidneys.
    c. brain.
    d. superior vena cava.

15. The weakening and ballooning of vein walls is called a(n)
    a. embolism.
    b. thrombus.
    c. myocardial infarction.
    d. aneurysm.

16. When a blood vessel is severed, the damaged epithelial tissue that lines the vessel would be
    a. mesothelium.
    b. simple columnar.
    c. endothelium.
    d. simple cuboidal.

17. When the quantity of epinephrine is increased
    a. the blood volume will increase
    b. the heart will slow down reflexly
    c. the heart will speed up reflexly
    d. the velocity of blood will decrease

18. In hepatic portal circulation, blood is eventually returned to the inferior vena cava through the
    a. superior mesenteric vein.
    b. portal vein.
    c. hepatic artery.
    d. hepatic veins.

19. Which of the following are involved in pulmonary circulation?
    a. superior vena cava, right atrium, and left ventricle
    b. inferior vena cava, right atrium, and left ventricle
    c. right ventricle, pulmonary artery, and left atrium
    d. left ventricle, aorta, and inferior vena cava

20. If a thrombus in the left common iliac vein becomes dislodged, into which arteriole system would it first find its way?
    a. brain
    b. kidneys
    c. lungs
    d. left arm

B. *Matching Questions.*   Each of the phrases in COLUMN B refers to a word or phrase in COLUMN A. Insert the letter of the word or phrase from COLUMN B that best describes it. Some words or phrases may be used more than once or not at all.

Column A		*Column B*	
1. ___	testes	**a.**	spermatic
2. ___	thigh	**b.**	splenic
3. ___	kidney	**c.**	phrenic
4. ___	spleen	**d.**	renal
5. ___	stomach	**e.**	gastric
6. ___	face	**f.**	posterior cerebral
7. ___	occipital lobe of cerebrum	**g.**	plantar
		h.	lumbar
8. ___	diaphragm	**i.**	internal maxillary
9. ___	sole of foot	**j.**	femoral
10. ___	abdominal wall		

Column A		*Column B*	
1. ___	abdominal aorta	**a.**	ulnar artery
2. ___	celiac artery	**b.**	posterior interventricular artery
3. ___	left coronary artery	**c.**	transverse sinuses
		d.	anterior tibial artery
4. ___	coronary sinus	**e.**	inferior mesenteric artery
5. ___	internal jugular veins	**f.**	hepatic artery
6. ___	right coronary artery	**g.**	median cubital vein
		h.	anterior interventricular artery
7. ___	brachial artery	**i.**	great and small cardiac veins
8. ___	popliteal artery	**j.**	superior mesenteric vein
9. ___	basilic vein		
10. ___	portal vein		

C. *True-False.*   Place a *T* or *F* in the space provided.

___ 1. Veins tend to have thick walls of elastic fibers and smooth muscles.

___ 2. The highest velocity blood flow occurs in the capillaries.

___ 3. Varicosities in the veins that lie in the wall of the rectum are called hemorrhoids.

___ **4.** Blood pressure is increased if either the flow or resistance is increased.

___ **5.** Blood flow is slower in the veins than in any other type of vessel.

___ **6.** If a blood vessel is stimulated by a vasoconstrictor nerve, the peripheral resistance increases.

___ **7.** The middle layer (tunica media) of arterioles contains large amounts of elastic fibers.

___ **8.** The right side of the heart is concerned with pulmonary circulation.

___ **9.** Blood entering the left atrium of the heart comes directly from the inferior vena cava and the superior vena cava.

___ **10.** The two circulatory paths in the body are called the pulmonary circulation (lesser circulation) and the systemic circulation (greater circulation).

___ **11.** The epithelial linings of the great vessels are continuous with the endocardium.

___ **12.** When vessels communicate with one another, they anastomose; an anastomosis is a passageway or connection between two vessels.

___ **13.** The spermatic and ovarian arteries are branches of the renal artery.

___ **14.** The unique feature of the portal system is that blood from the digestive tract is detoured through the liver instead of being returned directly to the inferior vena cava.

___ **15.** Umbilical veins carry oxygenated blood.

Answer Sheet—Chapter 14

Exercise 14.1

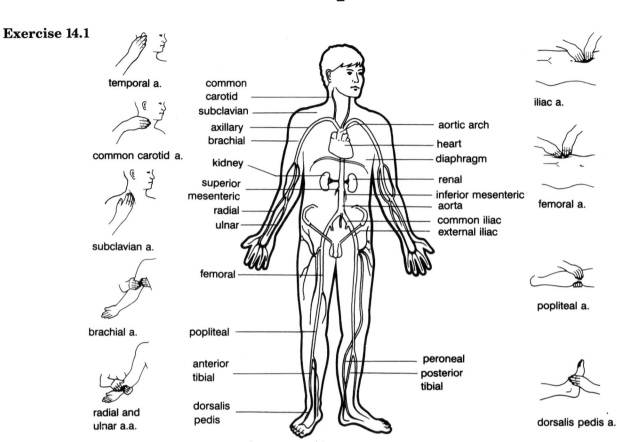

temporal a.

common carotid a.

subclavian a.

brachial a.

radial and ulnar a.a.

common carotid
subclavian
axillary
brachial
kidney
superior mesenteric
radial
ulnar

femoral

popliteal

anterior tibial

dorsalis pedis

aortic arch
heart
diaphragm
renal
inferior mesenteric
aorta
common iliac
external iliac

peroneal
posterior tibial

iliac a.

femoral a.

popliteal a.

dorsalis pedis a.

Figure 14.1   Major arteries and pressure points of the body.

Exercise 14.2

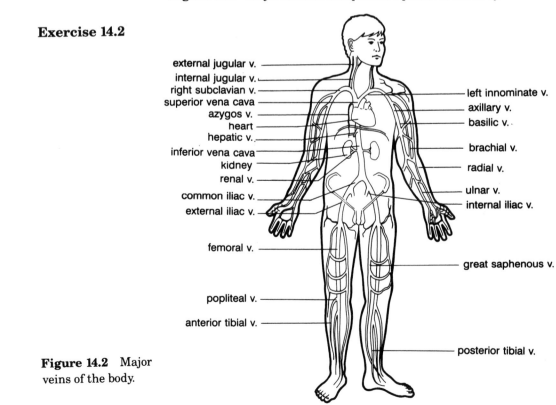

external jugular v.
internal jugular v.
right subclavian v.
superior vena cava
azygos v.
heart
hepatic v.
inferior vena cava
kidney
renal v.
common iliac v.
external iliac v.

femoral v.

popliteal v.

anterior tibial v.

left innominate v.
axillary v.
basilic v.

brachial v.

radial v.

ulnar v.
internal iliac v.

great saphenous v.

posterior tibial v.

Figure 14.2   Major veins of the body.

Exercise 14.3

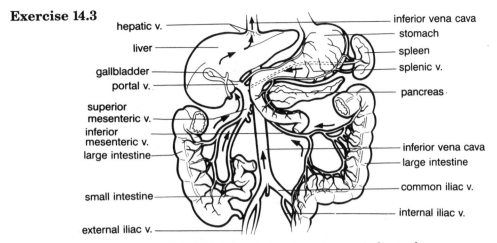

Figure 14.3 The hepatic portal system. Arrows indicate the direction of flow in all abdominal organs, through the liver, to the superior vena cava.

Exercise 14.4

PULMONARY CIRCULATION

The <u>pulmonary</u> circulation carries <u>blood</u> from the <u>heart</u> to the <u>lungs</u> and back to the heart again. The blood <u>leaves</u> the heart by way of the large pulmonary <u>artery</u> and <u>returns</u> through the pulmonary <u>veins</u>.

Originating in the <u>right</u> ventricle, the pulmonary <u>artery</u> passes upward and backward for a short distance and then <u>divides</u> into <u>right</u> and <u>left</u> branches that enter the <u>lungs</u>. Within each lung the artery <u>branches</u> widely, and each branch closely <u>follows</u> the subdivisions of the <u>bronchi</u>. The terminal <u>arterioles</u> open into the <u>capillary</u> networks that approximate the <u>aveolar</u> air sacs.

The pulmonary <u>veins</u> entering the <u>left</u> atrium have <u>two</u> branches from each <u>lung</u>. Their smallest tributaries arise as <u>venules</u> from the <u>capillary</u> networks of the alveolar air sacs.

Of paramount importance here is the <u>condition</u> of the blood in these vessels. As we have seen, blood <u>vessels</u> are <u>named</u> according to the <u>direction</u> of flow to or from the heart—not in terms of what they <u>carry</u>. Generally, a vessel leaving the heart is called an <u>artery</u> and carries <u>oxygenated</u> blood (red); a vessel <u>returning</u> to the heart is called a vein and carries <u>carbon dioxide</u>-laden blood (blue). However, in the pulmonary circulation only, the pulmonary <u>artery</u> carries <u>carbon dioxide</u>-laden blood <u>away</u> from the heart (to the lungs), and the pulmonary <u>vein</u> carries <u>oxygenated</u> blood <u>to</u> the heart (from the lungs).

Test Items

A. 1.d, 2.a, 3.b, 4.b, 5.d, 6.c, 7.a, 8.a, 9.b, 10.c, 11.d, 12.a, 13.c, 14.c, 15.d, 16.c, 17.c, 18.d, 19.c, 20.c.

B. 1.a, 2.j, 3.d, 4.b, 5.e, 6.i, 7.f, 8.c, 9.g, 10.h.
1.e, 2.f, 3.h, 4.i, 5.c, 6.b, 7.a, 8.d, 9.g, 10.j.

C. 1.F, 2.F, 3.T, 4.T, 5.F, 6.F, 7.T, 8.T, 9.F, 10.T, 11.T, 12.T, 13.F, 14.T, 15.T.

Circulation

Across

3 high blood pressure

5 pertaining to the arm

6 a hollow muscular tube that carries blood

8 from capillary to capillary

9 a vessel that carries blood to the heart

10 a vessel that carries blood away from the heart

12 a single-celled vessel that connects arterioles to venules

15 a sac-like enlargement of a blood vessel

16 thickening and hardening of the arteries

18 containing oxygen

19 pertaining to the lungs

20 pertaining to blood vessels

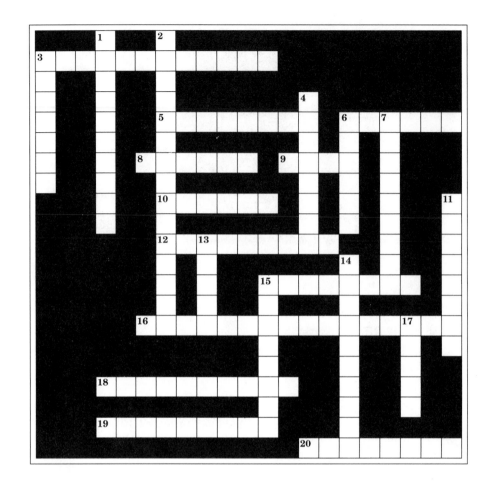

Down

1 bleeding

2 blood supply to the brain

3 pertaining to the liver

4 a sac-like enlargement

6 a small vein

7 superficial veins of the leg

11 abnormally swollen veins

13 rhythmic contraction of an artery

14 a small artery

15 a stroke

17 acute failure of blood circulation

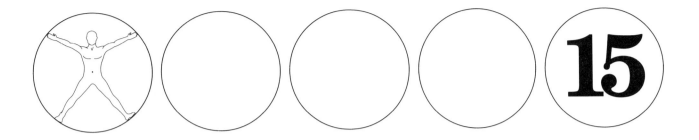

The Lymphatic System

I. CHAPTER SYNOPSIS

The lymphatic system is composed of lymph, lymph vessels, a series of small masses of lymphoid tissue called lymph nodes, and three organs—the tonsils, the thymus, and the spleen. Other principal areas of study include the role of lymphatic organs in antibody production, the immune response, allergic reactions, the basis for rejection of transplants, immunosuppressive techniques, implantation, and autoimmune diseases.

The lymphatic system is a one-way collecting system that gathers and drains filtered fluid and cellular constituents that accumulate in the spaces between the cells. The larger lymphatic vessels drain into veins, which return the lymph to the blood circulation.

Lymphatic capillaries resemble blood capillaries in structure. Both consist of a single layer of endothelial tissue. The major difference is that the lymphatic capillary has a closed terminal end; lymph fluid is absorbed from the tissue spaces through the endothelial membrane. Most of the tissues of the body, with the notable exception of the central nervous system, are drained by lymphatic vessels.

Larger lymphatic vessels drain the capillary network. The walls of these vessels resemble the walls of veins in structure. The muscle fibers in both the middle and outer layers are longitudinal and oblique, and therefore these larger vessels are contractile; lymphatic capillaries are not.

Lymphatic tissue filters and removes bacteria. Along the course of the lymphatic vessels are small bodies of lymphatic tissue called lymph nodes. These usually are oval and are commonly referred to as microkidneys.

II. OBJECTIVES

After reading the chapter, the student should be able to:

- Describe and identify the anatomical arrangement of a capillary bed.
- Compare the structure of a vein and lymph vessel.
- Identify the major vessels of the lymphatic system.
- Locate the major clusters of lymph nodes in the body.
- Explain how a superficial lymph node protects the body from infection.
- Explain tissue drainage.
- Describe edema and explain a number of ways it can occur.

III. IMPORTANT TERMS

Using your textbook, define the following terms:

capillary (kap'-ah-ler-ee) _____

_____

chyle (kile) _____

_____

cisterna (sis-tur'-nah) _____

_____

duct (dukt) _____

_____

elephantiasis (el-ah-fahn-ti'-ah-sis) _____

_____

endothelial (en-do-thee'-lee-ul) _____

_____

extracellular (ek-strah-sel'-yah-lur) _____

_____

intercellular (in-tur-sel'-yah-lur) _____

_____

intercostal (in-tur-kos'-tul) _____

_____

lacteal (lack'-tee-ul) _____

_____

lymph (limf) _____

_____

mastectomy (ma-stek'-tah-mee) _____

_____

node (node) _____

_____

splenomegaly (splen-o-meg'-ah-lee) _____

_____

thoracic (tho-ras'-ik) _____

_____

IV. EXERCISES

Complete the following exercises in the order given. A precise set of terms and diagrams has been chosen to describe the lymphatic system.

Exercise 15.1

Labeling. Write the name of the structure in the space provided. Color the vessels differently from the surrounding tissue.

Key:

arterial	endothelial	venous
blind	intercellular	
capillary	tissue	

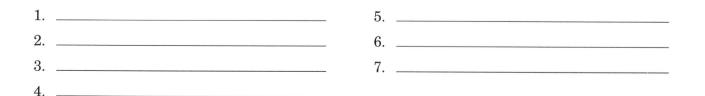

Figure 15.1 Diagram of a capillary bed, showing how materials diffuse between arterial capillaries and venous capillaries. Materials that are trapped in the intercellular tissue spaces are collected by lymphatic capillaries and returned to the blood system.

1. _____   5. _____

2. _____   6. _____

3. _____   7. _____

4. _____

Exercise 15.2

Labeling.   Write the name of the structure in the space provided. Color the organs and vessels differently.

Key:
axillary
capillaries
cisterna
cubital
duct
heart
inguinal
lymph
palmar
parotid
plantar
popliteal
thoracic

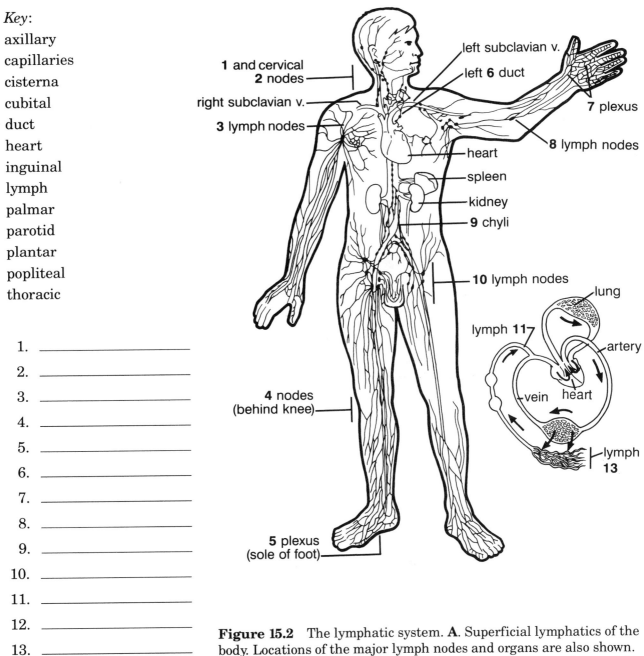

1 and cervical 2 nodes
right subclavian v.
3 lymph nodes
left subclavian v.
left 6 duct
7 plexus
8 lymph nodes
heart
spleen
kidney
9 chyli
10 lymph nodes
lung
lymph 11
artery
vein   heart
lymph 13
4 nodes (behind knee)
5 plexus (sole of foot)

1. _____
2. _____
3. _____
4. _____
5. _____
6. _____
7. _____
8. _____
9. _____
10. _____
11. _____
12. _____
13. _____

Figure 15.2   The lymphatic system. **A.** Superficial lymphatics of the body. Locations of the major lymph nodes and organs are also shown. Shaded area indicates the segment of the upper right quadrant of the body that drains into the right lymphatic duct. The remaining area of the body is drained by the left lymphatic (thoracic) duct. **B.** Diagrammatic representation of the lymphatic system, showing its connection with the general circulatory system.

Exercise 15.3

Labeling. Write the name of the structure in the space provided. Color the organs and vessels differently as in Figure 15.2.

Key:

brachiocephalic
cisterna
heart
intercostal
intestine
jugular
kidney
liver
nodes
pulmonary
spleen
subclavian
thoracic
vena cava
vessels

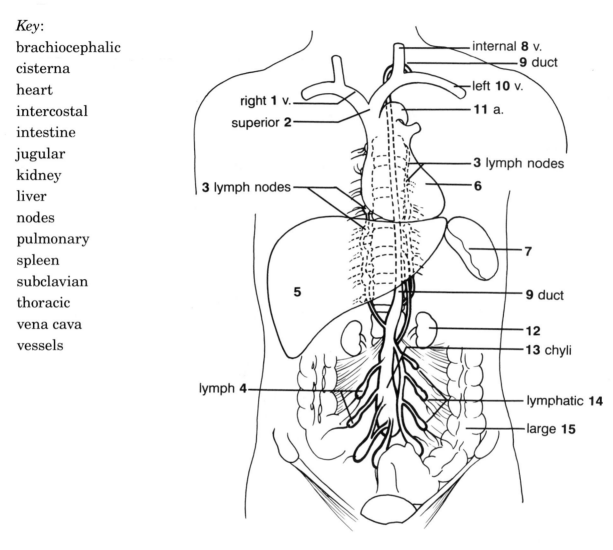

Figure 15.3 Lymphatic drainage to the cisterna chyli and thoracic duct.

1. _____
2. _____
3. _____
4. _____
5. _____
6. _____
7. _____
8. _____
9. _____
10. _____
11. _____
12. _____
13. _____
14. _____
15. _____

V. TEST ITEMS

A. *Multiple Choice.* There is only one answer that is either correct or most appropriate. Circle the answer that corresponds to the question.

1. Lymph flow is aided by all the following factors except
   a. breathing.
   b. muscular contraction.
   c. protein osmotic pressure.
   d. valves in lymph vessels.

2. Lymph
   a. does not always reach the bloodstream.
   b. contains large amounts of protein.
   c. carries a lower waste content than the blood.
   d. flows against gravity in the lymphatics.

3. The lymphatic system returns interstitial fluid to the
   a. liver.
   b. blood.
   c. CSF.
   d. intestinal lumen.

4. Which one of the following would not be considered part of the lymphatic system?
   a. thymus
   b. lymph node
   c. tonsil
   d. heart

5. The thoracic duct receives lymph from a dilated vessel called the
   a. cisterna chyli.
   b. lacteal.
   c. lymph capillary.
   d. right lymphatic duct.

6. The thoracic duct delivers lymph into the
   a. large lymphatics.
   b. lymph capillaries.
   c. right lymphatic duct.
   d. left subclavian vein.

7. A large organ structurally similar to a lymph node but designed to filter blood is the
   a. thymus.
   b. spleen.
   c. Peyer's patch.
   d. pharyngeal tonsil.

8. Lymphatics and veins are similar in that both
   a. contain valves.
   b. carry oxygenated blood.
   c. carry deoxygenated blood.
   d. carry blood away from the heart.

9. Lymph
   a. flows at approximately the same rate as blood in the veins.
   b. has the same composition as blood plasma.
   c. is tissue fluid that enters the lymph capillaries.
   d. normally contains large numbers of red blood cells.

10. Lymph glands
    a. have only one afferent vessel.
    b. filter blood.
    c. produce large numbers of lymphocytes.
    d. in the extremities are located primarily in the hands and feet.

11. The spleen
    a. is the primary reservoir for lymph.
    b. produces large numbers of granulocytes during adulthood.
    c. destroys wornout red blood cells.
    d. is necessary for life.

12. The lymphatic system and the venous system have which of the following properties in common?
    a. They move fluid toward the heart.
    b. The flow of fluid in the vessels is aided by skeletal muscle contractions.
    c. They contain red blood cells.
    d. The walls of the vessels are impermeable to proteins.

13. Why should lymph in the thoracic duct have a much higher lymphocyte count than lymph in peripheral lymph spaces? The lymphocytes
    a. have just circulated through the tissues and are carrying waste material.
    b. have multiplied by mitotic division in the lymph stream.
    c. have increased in numbers because this lymph has just passed through numerous lymph nodes.
    d. have not as yet been filtered out by passage through the lymph nodes.

14. Lymphatic tissue is present in organs other than lymph nodes. Which one of the following is *not* a lymphatic organ?
    a. lymphoid nodules and tonsils   c. liver
    b. spleen   d. thymus

15. What is the function of the lymphatic capillaries in the villi of the intestine? They absorb
    a. sugars.   c. proteins.
    b. fats.   d. water.

16. An abnormal accumulation of fluid in tissue spaces is called
    a. diuresis.   c. edema.
    b. bursitis.   d. anemia.

17. What do we call the lymph that is carrying absorbed fat and has a milk-white appearance?
    a. lacteal   c. chyle
    b. bile   d. synovial fluid

18. The lymph capillary within the villus of the intestine is called a
    a. thoracic duct.   c. chyle duct.
    b. lacteal.   d. milk duct.

19. Which of the following tonsils are found around the opening of the digestive and respiratory systems?
    a. palatine   c. lingual
    b. pharyngeal   d. all of the above

20. The major lymph vessel in the body is
    a. the thoracic duct.   c. the intestinal trunk.
    b. the right lymphatic duct. d. none of the above

B. *Matching Questions.*   Each of the phrases in COLUMN B refers to a word or phrase in COLUMN A. Insert the letter of the word or phrase from COLUMN B that best describes it. Some words or phrases may be used more than once or not at all.

Column A	Column B
1. ___ edema	**a.** substance produced in response to an antigen
2. ___ adenitis	**b.** skin eruption
3. ___ antigen	**c.** enlargement of a gland
4. ___ immune	**d.** a benign lymph tumor
5. ___ effusion	**e.** filaria worm
6. ___ lymphosarcoma	**f.** escape of fluid from lymphatics
7. ___ adenopathy	**g.** resistance to a disease
8. ___ antibody	**h.** swelling of tissues
9. ___ hives	**i.** inflammation of adenoids
10. ___ elephantiasis	**j.** a malignant lymph tumor
	k. foreign substance

Column A	Column B
1. ___ lacteal	**a.** left lymphatic duct
2. ___ thoracic duct	**b.** lymph nodule
3. ___ palatine	**c.** behind the knees
4. ___ thymosine	**d.** enlargement of lymph nodes
5. ___ germinal center	**e.** lymphatic capillary
6. ___ cisterna chyli	**f.** inguinal region
7. ___ popliteal	**g.** removal of breast tissue
8. ___ Hodgkin's disease	**h.** tonsil
9. ___ mastectomy	**i.** enlarged spleen
10. ___ splenomegaly	**j.** thymus hormone
	k. intestinal drainage

C. *True-False.*   Place a *T* or *F* in the space provided.

_____ **1.** The excessive accumulation of lymph in tissue spaces is referred to as edema.

_____ **2.** The combination of antibodies with antigens on a bacterial cell wall kills the bacterium.

_____ **3.** The initial interaction of antigens with the body's defense system leading to the formation of antibodies occurs in cells located in the lymph nodes.

_____ **4.** The body produces larger quantities of antigens the first time it encounters a foreign antigen than upon subsequent encounters with the same antigen.

_____ **5.** The phagocytosis of bacteria by white blood cells is increased when the bacterial surface antigens have combined with antibodies.

_____ **6.** The increased blood flow accompanying inflammation ensures an adequate supply of leukocytes, which play a crucial role in the immune response.

_____ **7.** Interferon is particularly effective against bacterial invasions.

_____ **8.** Specific interferon molecules attack only those foreign substances that "fit" their protein structure.

_____ **9.** The interferon system reacts rapidly, with increased synthesis beginning within hours of the onset of the infection.

_____ **10.** Antibodies cannot enter intact human cells, but interferon can.

_____ **11.** In general, the B cells and the T cells serve identical functions.

_____ **12.** Antigens are identical proteins except for a relatively small number of amino acids occupying the first position in the chains.

_____ **13.** The transfer of actively formed antibodies from one person (or animal) to another confers a resistance known as active immunity.

_____ **14.** Persons with type AB blood have neither anti-A nor anti-B antibodies.

_____ **15.** Lymphatic vessels have valves.

_____ **16.** The main thoracic duct empties into the left brachiocephalic vein.

_____ **17.** Lymphatic tissue is present only in lymph nodes and nodules that are present only in certain special regions of the body such as the axillary region.

_____ **18.** Mast cells and basophils are the main sources of histamine—a vasodiatory substance.

_____ **19.** Any organic molecule, usually large, that has been recognized as foreign by the body's defense system can function as an antigen.

_____ **20.** Lymph is returned to the heart by a system of vessels structured like arteries.

Answer Sheet—Chapter 15

Exercise 15.1

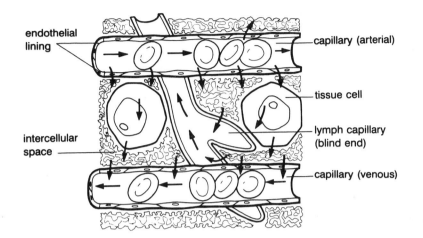

Figure 15.1   Diagram of a capillary bed, showing how materials diffuse between arterial capillaries and venous capillaries. Materials that are trapped in the intercellular tissue spaces are collected by lymphatic capillaries and returned to the blood system.

Exercise 15.2

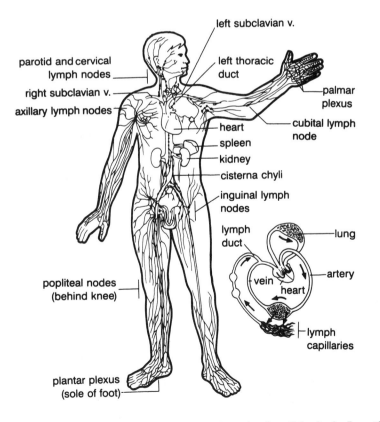

Figure 15.2   The lymphatic system. **A**. Superficial lymphatics of the body. Locations of the major lymph nodes and organs are also shown. Shaded area indicates the segment of the upper right quadrant of the body that drains into the right lymphatic duct. The remaining area of the body is drained by the left lymphatic (thoracic) duct. **B**. Diagrammatic representation of the lymphatic system, showing its connection with the general circulatory system.

Exercise 15.3

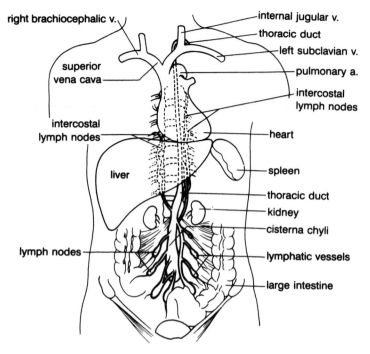

Figure 15.3 Lymphatic drainage to the cisterna chyli and thoracic duct.

Test Items

A. 1.c, 2.d, 3.b, 4.d, 5.a, 6.d, 7.b, 8.a, 9.c, 10.c, 11.c, 12.a, 13.c, 14.c, 15.b, 16.c, 17.c, 18.b, 19.d, 20.a.

B. 1.h, 2.i, 3.k, 4.g, 5.f, 6.j, 7.c, 8.a, 9.b, 10.e.
1.e, 2.a, 3.h, 4.j, 5.b, 6.k, 7.c, 8.d, 9.g, 10.i.

C. 1.T, 2.F, 3.T, 4.F, 5.T, 6.T, 7.F, 8.F, 9.T, 10.F, 11.F, 12.F, 13.F, 14.T, 15.T, 16.F, 17.F, 18.T, 19.T, 20.F.

The Lymphatic System

Across

1 milky fluids absorbed by the lacteals of the small intestine

5 the inner layer of cells of a vessel or organ

9 a hollow tubule

11 a tissue fluid that contains most of the components of blood

13 pertaining to the chest

15 a benign lymphatic tumor

16 a microscopic structure that contains lymphocytes

17 an enlarged spleen

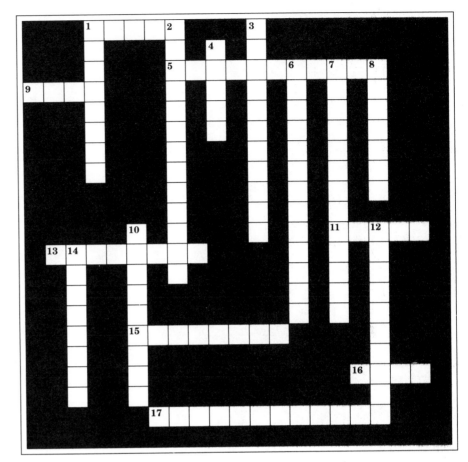

Down

1 a system of lymph vessels

2 enlarged tissue due to inadequate fluid drainage

3 between the ribs

4 excessive tissue fluid

6 space outside of a cell

7 space inside of a cell

8 a one-celled vessel of the lymphatic system

10 a one-celled vessel of the blood circulatory system

12 surgical removal of the mammary gland

14 a malignancy of the lymphatic system

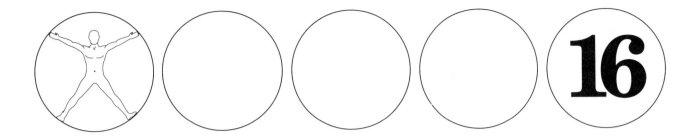

The Respiratory System

I. CHAPTER SYNOPSIS

The utilization of oxygen by the mitochondria during the process of oxidative phosphorylation provides the major mechanism for the synthesis of ATP by the cells of the body. The mitochondria are also the site of carbon dioxide production, the major metabolic waste product of metabolism. The lungs and circulatory system provide the mechanisms for delivering oxygen to the cells and removing carbon dioxide. During the course of evolution, specialized proteins (hemoglobin) evolved that increase the oxygen-carrying capacity of the blood. The conversion of carbon dioxide to carbonic acid in the red blood cells and its dissociation into bicarbonate and hydrogen ions provide a mechanism for regulating the acidity of body fluids by controlling the volume of expired carbon dioxide. The levels of carbon dioxide and oxygen in the blood are powerful regulatory agents for controlling the activity of both the respiratory and cardiovascular systems. Thus, this chapter examines the organization of the respiratory system, and the steps involved in respiration, ventilation, and exchange and transport of gases in the body.

II. OBJECTIVES

After reading the chapter, the student should be able to:

- Label the structures of the respiratory system.

- Identify two functions of the cilia and mucous cells.

- Define respiration, ventilation, inspiration, and expiration.

- Describe the vagal Breuer control of breathing.

- Explain the oxygen-associated chemoreceptor mechanism for the control of ventilation.

- Explain the process of gaseous exchange in the alveoli and in the tissue.

III. IMPORTANT TERMS

Using your textbook, define the following terms:

alveolar (al-vee'-ah-lahr) _____

_____

aspiration (as-pah-ray'-shun) _____

_____

atrium (ay'-tree-um) _____

_____

bronchiole (brong'-kee-ol) _____

_____

bronchus (brong'-kus) _____

_____

concha (kong'-kah) _____

_____

cortex (kor'-teks) _____

_____

dyspnea (disp-nee'-ah) _____

_____

emphysema (emp-fah-zee'-mah) _____

_____

epiglottis (ep-ah-glot'-is) _____

_____

expiration (ek-spah-ray'-shun) _____

_____

glottis (glot'-is) _____

_____

hyaline (hi'-ah-lin) _____

_____

inspiration (int-spah-ray'-shun) _____

_____

intrapulmonary (in-trah-pul'-mah-ner-ee) _____

_____

intrathoracic (in-trah-tho-ras'-ik) _____

_____

larynx (lar'-inks) _____

_____

lingual (ling'-gwul) _____

_____

nasopharynx (nay-zo-far'-inks) _____

_____

oropharynx (or-o-far'-inks) _____

_____

phrenic (fren'-ik) _____

_____

pneumotaxic (new-mo-taks'-ik) _____

_____

respiration (res-pah-ray'-shun) _____

_____

tonsil (ton'-sil) _____

_____

vagus (vay'-gus) _____

_____

ventilation (vent-il-ay'-shun) _____

_____

IV. EXERCISES

Complete the following exercises in the order given. A precise set of terms and diagrams has been chosen to describe the respiratory system.

Exercise 16.1

Labeling. Write the name of the structure in the space provided. Color the nasal and oral passages differently.

Key:

concha

epiglottis

esophagus

glottis

larynx

lingual

mandible

nasopharynx

oral

oropharynx

thyroid

tongue

tonsil

trachea

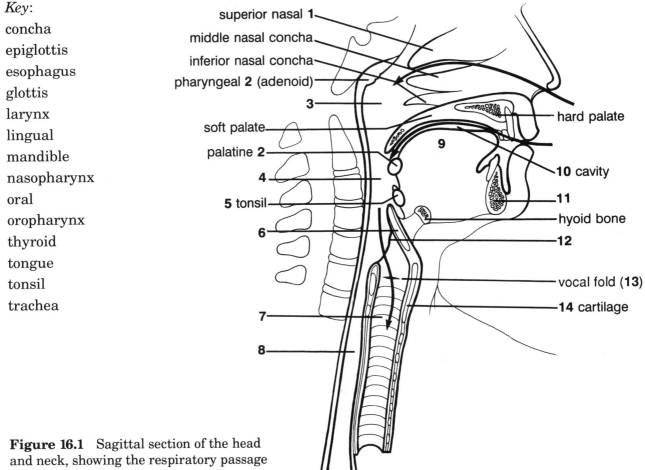

superior nasal **1**
middle nasal concha
inferior nasal concha
pharyngeal **2** (adenoid)
3
soft palate
palatine **2**
4
5 tonsil
6
7
8

hard palate
9
10 cavity
11
hyoid bone
12
vocal fold (**13**)
14 cartilage

Figure 16.1 Sagittal section of the head and neck, showing the respiratory passage down to the bifurcation of the trachea.

1. _____   8. _____

2. _____   9. _____

3. _____   10. _____

4. _____   11. _____

5. _____   12. _____

6. _____   13. _____

7. _____   14. _____

Exercise 16.2

Labeling.   Write the name of the structure in the space provided. Color the respiratory tract and the lung mass differently.

Key:
alveolar
alveoli
atrium
bronchiole
bronchioles
bronchus
capillary
duct
glottis
respiratory
sacs
trachea

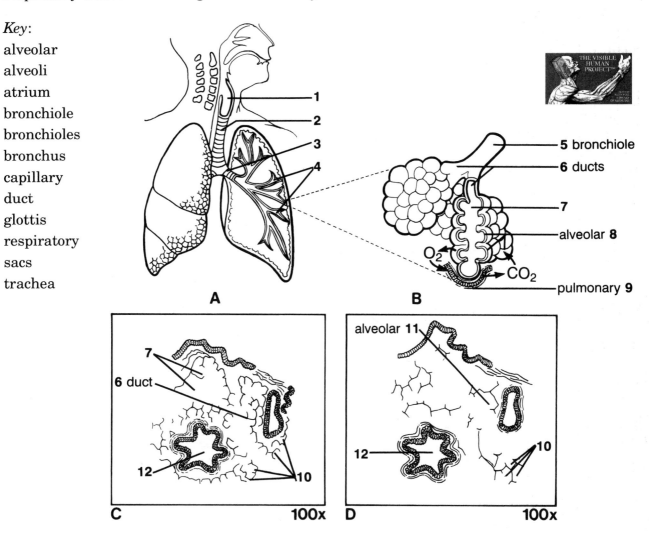

Figure 16.2   Internal structure of the lungs. **A.** Relationship of the lungs to the head and neck. **B.** External and internal appearance of a lung lobule, showing the atrium and alveolar sacs. **C.** Section of a lung, magnified 100 times. **D.** Section of a similar lung with emphysema. Note the decreased number of alveolar sacs and consequent diminishing in the gas exchange area of lung tissue.

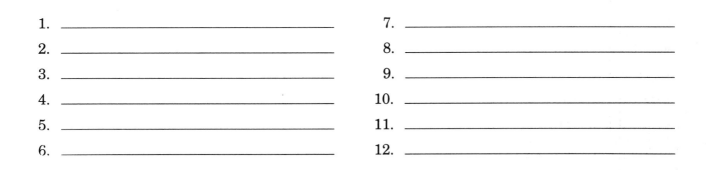

1. _____
2. _____
3. _____
4. _____
5. _____
6. _____

7. _____
8. _____
9. _____
10. _____
11. _____
12. _____

Exercise 16.3

Labeling. Write the name of the structure in the space provided. Color the various organs and tissues differently.

Key:
aortic
CO₂
carotid
cerebrum
cortex
diaphragm
expiratory
glossopharyngeal
inspiratory
intercostal
medulla
phrenic
pneumotaxic
stretch
vagus
voluntary

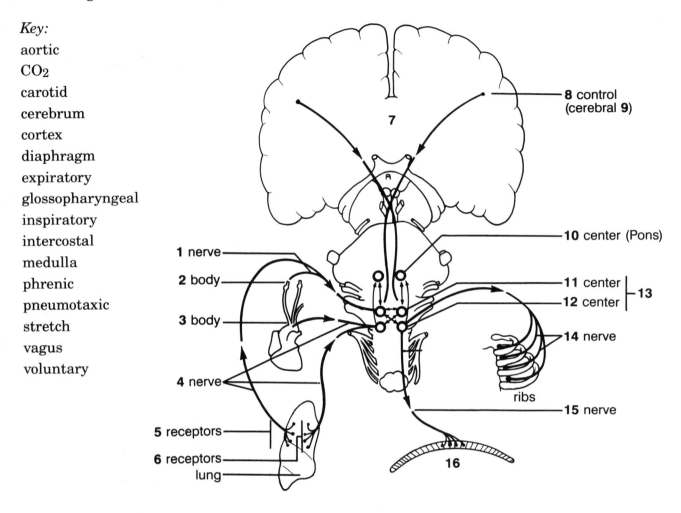

Figure 16.3   Chemical and nervous control of respiration. Arrows indicate direction of nerve impulses.

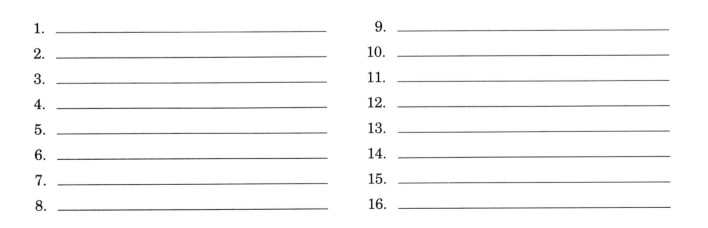

1. _____
2. _____
3. _____
4. _____
5. _____
6. _____
7. _____
8. _____

9. _____
10. _____
11. _____
12. _____
13. _____
14. _____
15. _____
16. _____

Exercise 16.4

Completion.   Write the correct terms in the spaces provided.

Key:

alveoli

aorta

arterial

arteries

blood

carbon dioxide

carbon dioxide

cellular

lungs

oxygen

oxygen

returned

returns

tissues

veins

vena cava

venous

GASEOUS EXCHANGE

_____ gas exchange takes place in all body _____ ; cells take up _____ from the _____ and add _____ to it. When this occurs, blood becomes _____. As we have seen, this carbon dioxide-rich blood then _____ to the heart, entering through the _____. From the heart, venous blood is pumped through the short, paired pulmonary _____ to the _____, where these arteries branch into extensive capillary networks spread over the _____ . Pulmonary gas exchange takes place here: _____ leaves the blood and _____ enters. Thus blood becomes _____. This oxygen-rich blood now is _____ to the heart by the pulmonary _____, and this blood in turn is redistributed throughout the body by the _____ .

Key:

air

capillary

cell

determined

direction

greater

into

pressure

saturated

venous

vessel

The _____ in which the gases move is _____ by the prevailing _____ gradients, or tension gradients, between blood _____ and organ _____. Specifically, atmospheric _____ in the alveoli contains only a little carbon dioxide, but the _____ blood that flows into the lungs from the body is virtually _____ with the gas. The pressure gradient of carbon dioxide is _____ in the _____ ; therefore, carbon dioxide diffuses _____ the alveolus.

Key:

blood

capillary

from

high

into

into

low

low

oxygen

oxygen

oxygen

pressure

respiratory

reverse

tension

tissue

waste

Predictably, the pressure pattern is the _____ for oxygen. Blood flowing into the lungs from the body is _____-poor. The air in the alveoli contains a maximal content of oxygen; therefore oxygen diffuses _____ the _____ from the alveolus to satisfy the increased alveolar _____ gradient.

Just as in the lungs, cellular gas exchange in the other body tissues is governed by _____ gradients. Cells continuously use up _____ in internal respiration, and the tension of this gas in cells is therefore _____ . The arterial tension is high, however, and the _____ diffuses _____ the blood _____ the tissue cells. At the same time, since _____ carbon dioxide is produced as a _____ product of cellular metabolism, the carbon dioxide tension within the tissue cells is _____ . Arterial blood has a _____ carbon dioxide tension, and the gas therefore diffuses from the _____ cells into the _____ .

V. TEST ITEMS

A. *Multiple Choice.* There is only one answer that is either correct or most appropriate. Circle the answer that corresponds to the question.

1. The space between the pleura of the lungs that extends from the breastbone to the backbone is the
   a. cranium.
   b. mediastinum.
   c. hypogastric region.
   d. epigastric region.

2. A rise in the carbonic acid content of the blood will have what effect on respiration?
   a. The respiratory rate increases and the breathing becomes stronger and deeper.
   b. The respiratory rate decreases and the breathing becomes shallow and weaker.
   c. The chemoreceptors are depressed.
   d. The pressoreceptors are depressed.

3. Which of the following statements correctly describes the conditions of the lungs prior to birth?
   a. They are collapsed.
   b. They totally fill the thoracic cavity.
   c. Amniotic fluid is breathed in and out of the lungs.
   d. Intrapleural pressure is greater than atmospheric pressure.

4. Which portion of the pharynx serves solely as a respiratory passage-way?
   a. nasal pharynx
   b. oral pharynx
   c. laryngeal pharynx

5. Too rapid decompression after exposure to high atmospheric pressure may cause gas bubbles to form in the blood and tissues; a dangerous condition called the bends. What is the gas that causes these symptoms?
   a. carbon dioxide   c. ammonia
   b. oxygen   d. nitrogen

6. Why is the inhalation of carbon monoxide so dangerous?
   a. It causes an increase in the respiratory rate.
   b. It causes a rise in blood pressure.
   c. Because hemoglobin has a much greater attraction for CO than it does for oxygen.
   d. It causes a breakdown of the surface active agent in the alveoli.

7. Upon opening the chest, which structure would you see first?
   a. parietal pleura   c. visceral pleura
   b. secondary bronchi   d. pleural cavity

8. What would you call the cavity that lies between the lungs? The heart lies in this cavity and the diaphragm forms its floor.
   a. abdominal
   b. pleural cavity
   c. pericardial cavity
   d. peritoneal cavity

9. Where are the inspiratory and expiratory centers located? In the
   a. lungs.   c. cerebellum.
   b. intercostal muscles.   d. medulla.

10. In what structures are oxygen from the air and carbon dioxide from the blood exchanged? In the
    a. lacteals of the villi.
    b. pleura.
    c. bronchioles.
    d. alveoli.

11. Name the cartilaginous structure that contains the vocal folds:
    a. pharynx   c. epiglottis
    b. larynx   d. bronchus

12. The greatest amount, about 75 percent of carbon dioxide in the blood, is carried by
    a. bicarbonate ions, HCO_3.
    b. hemoglobin.
    c. blood plasma.
    d. carbonic anhydrase.

13. The reaction of carbon dioxide with water is a relatively slow process, yet this reaction occurs within the red cells in a fraction of a second. What is responsible for speeding up this reaction?
    a. cytochrome oxidase
    b. ATP
    c. phosphorylase
    d. carbonic anhydrase

14. If a person inhales the maximum amount of air and then exhales as much as possible the total exchange represents his
    a. tidal air flow.
    b. residual air volume.
    c. inspiratory reserve.
    d. vital capacity.

15. If we cut the phrenic nerves in an experimental animal, what happens?
    a. The blood pressure rises.
    b. The heart rate increases.
    c. Peristalsis is inhibited.
    d. The diaphragm stops moving.

16. When you take a very deep breath of air just before diving into water, this volume of air is referred to as
    a. tidal volume.
    b. expiratory reserve volume.
    c. vital capacity.
    d. inspiratory reserve volume.

17. Which statement about the Hering-Breuer reflex is not true?
    a. When the lungs are filled with air the inspiratory center in the medulla is stimulated.
    b. When the lungs are deflated stretch receptors are no longer stimulated.
    c. When the lungs are filled with air the expiratory center in the medulla is stimulated.
    d. When the lungs are inflated stretch receptors are stimulated.

18. A spasmodic contraction of the diaphragm followed by a spasmodic closure of the glottis that results in a sharp inspiratory sound best describes a
    a. hiccough.
    b. cough.
    c. sneeze.
    d. sigh.

19. Which sequence of events best describes inspiration?
    a. contraction of diaphragm and intercostals, decrease in intrathoracic pressure, decrease in intrapulmonic pressure
    b. relaxation of diaphragm and intercostals, decrease in intrathoracic pressure, decrease in intrapulmonic pressure
    c. contraction of diaphragm and intercostals, increase in intrathoracic pressure, increase in intrapulmonic pressure
    d. relaxation of diaphragm and intercostals, increase in intrathoracic pressure, increase in intrapulmonic pressure

20. A cessation of respiration due to lack of stimulation of the respiratory center is called
    a. asphyxia.
    b. apnea.
    c. dyspnea.
    d. hyperpnea.

B. *Matching Questions.* Each of the phrases in COLUMN B refers to a word or phrase in COLUMN A. Insert the letter of the word or phrase from COLUMN B that best describes it. Some words may be used more than once or not at all.

Column A	*Column B*
1. ___ alveoli	**a.** an air space under a concha
2. ___ ethmoid	**b.** a bone of the nasal cavity
3. ___ adenoids	**c.** a large cavity within a maxillary bone
4. ___ mediastinum	**d.** a common passage for air and food
5. ___ pharynx	**e.** the sites of gas exchange within the lungs
6. ___ tonsils	**f.** the slight dilation of each nasal cavity just inside the nostril
7. ___ hilum	
8. ___ columnar epithelium	**g.** hypertrophied pharyngeal tonsils
	h. lymphoid tissue
9. ___ meatus	**i.** produces sounds
10. ___ larynx	**j.** the largest cartilage of the lung
11. ___ mucus	**k.** separates each lung
12. ___ paranasal sinus	**l.** the main surface depression of the lung
13. ___ diaphragm	**m.** contain cilia
14. ___ thyroid	**n.** is constantly moved by cilia to the pharynx
15. ___ vestibule	**o.** the principal muscle of respiration

		Column B	
Column A			
1. ___	apnea	a.	sounds heard in the lungs that resemble bubbling
2. ___	diphtheria	b.	irregular breathing beginning with shallow breaths that increase in depth and rapidity then decrease and cease altogether
3. ___	hypoxia		
4. ___	orthopnea	c.	a temporary absence of respirations
5. ___	pneumothorax	d.	oxygen starvation
6. ___	bronchitis	e.	bacterial infection that causes the membranes of the pharynx and larynx to become enlarged and leathery
7. ___	pulmonary embolism		
		f.	presence of clot in a pulmonary arterial vessel that stops circulation to a part of the lungs
8. ___	rales		
9. ___	respirator	g.	reduction in oxygen supply to cells
10. ___	influenza	h.	normal quiet breathing
11. ___	dyspnea	i.	viral infection that causes inflammation of respiratory mucous membranes and fever
12. ___	eupnea		
13. ___	asphyxia	j.	presence of more then normal air in the pleural cavity
14. ___	Cheyne-Stokes respiration	k.	a collapsed lung or portion of a lung
		l.	inability to breathe in a horizontal position
15. ___	atelectasis	m.	labored or difficult breathing
		n.	a machine that can maintain artificial respiration
		o.	inflammation of the bronchi and bronchioles

C. *True-False.*   Place a *T* or *F* in the space provided.

___ **1.** The palatine tonsils are located on the posterior wall of the naso-pharynx.

___ **2.** During inspiration, contraction of the diaphragm and external intercostal muscles increases the size of the thoracic cavity.

___ **3.** Tidal volume is usually about 500 cc.

___ **4.** During expiration, the intrathoracic (intrapleural) pressure becomes positive.

___ **5.** During quiet breathing, expiration is brought about by the contraction of the external intercostal muscles.

___ **6.** Vital capacity is the maximum amount of air expired after a maximum inspiration.

___ **7.** Most of the CO_2 in the blood is carried by hemoglobin.

___ **8.** Irritation of the vagus nerve will result in deep, slow breathing.

___ **9.** An increase in the concentration of carbon dioxide in the blood increases ventilation by chemical stimulation of the respiratory center in the medulla.

___ **10.** Enlarged chest, degeneration of alveolar sacs, high levels of carbon dioxide in the blood, and development of fibrous connective tissue are all symptoms of pneumonia.

___ **11.** Gases move between the blood and lungs by the process of filtration.

___ **12.** Inflammation of the membrane around the lung is referred to as bronchitis.

___ **13.** Any sudden increase in arterial blood pressure will decrease the rate of respiration.

___ **14.** The top portion of the nose communicates with the pharynx through the internal nares.

___ **15.** The major portion of carbon dioxide is carried in the blood as the bicarbonate ion.

___ **16.** The pneumotaxic center is located in the medulla and assumes a role in the control of respirations.

___ **17.** If a person undergoes continued hypoventilation, the CO_2 will increase.

___ **18.** A spasmodic contraction of the expiratory muscles that forcefully expels air through the nose and mouth is called a cough.

___ **19.** The combination of external cardiac compression and exhaled air ventilation constitutes heart-lung resuscitation.

___ **20.** The term eupnea refers to cessation of breathing at the end of a normal expiration.

Answer Sheet—Chapter 16

Exercise 16.1

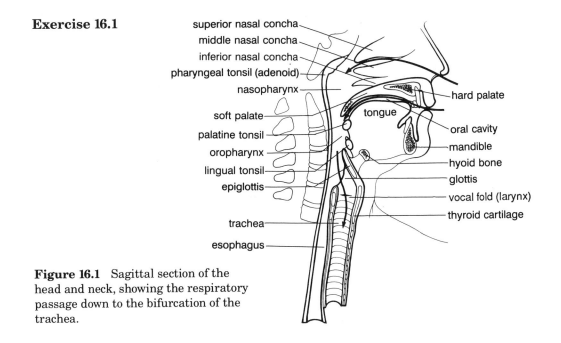

superior nasal concha
middle nasal concha
inferior nasal concha
pharyngeal tonsil (adenoid)
nasopharynx
soft palate
palatine tonsil
oropharynx
lingual tonsil
epiglottis
trachea
esophagus

hard palate
tongue
oral cavity
mandible
hyoid bone
glottis
vocal fold (larynx)
thyroid cartilage

Figure 16.1 Sagittal section of the head and neck, showing the respiratory passage down to the bifurcation of the trachea.

Exercise 16.2

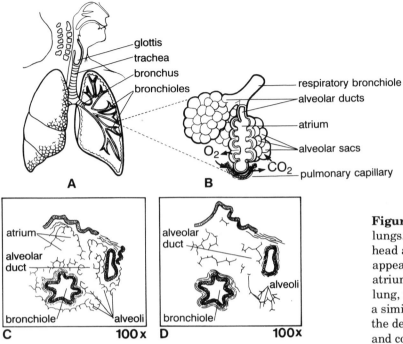

A

B

C   100x   D   100x

Figure 16.2   Internal structure of the lungs. **A**. Relationship of the lungs to the head and neck. **B**. External and internal appearance of a lung lobule, showing the atrium and alveolar sacs. **C**. Section of a lung, magnified 100 times. **D**. Section of a similar lung with emphysema. Note the decreased number of alveolar sacs and consequent diminishing in the gas exchange area of lung tissue.

Exercise 16.3

Figure 16.3   Chemical and nervous control of respiration. Arrows indicate direction of nerve impulses.

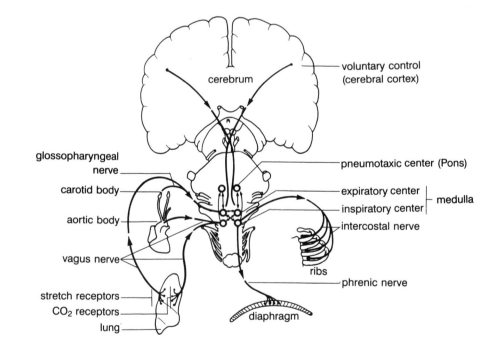

Exercise 16.4

GASEOUS EXCHANGE

<u>Cellular</u> gas exchange takes place in all body <u>tissues</u>; cells take up <u>oxygen</u> from the <u>blood</u> and add <u>carbon dioxide</u> to it. When this occurs, blood becomes <u>venous</u>. As we have seen, this carbon dioxide-rich blood then <u>returns</u> to the heart, entering through the <u>vena cava</u>. From the heart, venous blood is pumped through the short, paired pulmonary <u>arteries</u> to the <u>lungs</u>, where these arteries branch into extensive capillary networks spread over the <u>alveoli</u>. Pulmonary gas exchange takes place here: <u>carbon dioxide</u> leaves the blood and <u>oxygen</u> enters. Thus blood becomes <u>arterial</u>. This oxygen-rich blood now is <u>returned</u> to the heart by the pulmonary <u>veins</u>, and this blood in turn is redistributed throughout the body by the <u>aorta</u>.

The <u>direction</u> in which the gases move is <u>determined</u> by the prevailing <u>pressure</u> gradients, or tension gradients, between blood <u>vessel</u> and organ <u>cell</u>. Specifically, atmospheric <u>air</u> in the alveoli contains only a little carbon dioxide, but the <u>venous</u> blood that flows into the lungs from the body is virtually <u>saturated</u> with the gas. The pressure gradient of carbon dioxide is <u>greater</u> in the <u>capillary</u>; therefore, carbon dioxide diffuses <u>into</u> the alveolus.

Predictably, the pressure pattern is the <u>reverse</u> for oxygen. Blood flowing into the lungs from the body is <u>oxygen</u>-poor. The air in the alveoli contains a maximal content of oxygen; therefore oxygen diffuses <u>into</u> the <u>capillary</u> from the alveolus to satisfy the increased alveolar <u>pressure</u> gradient.

Just as in the lungs, cellular gas exchange in the other body tissues is governed by <u>tension</u> gradients. Cells continuously use up <u>oxygen</u> in internal respiration, and the tension of this gas in cells is therefore <u>low</u>. The arterial tension is high, however, and the <u>oxygen</u> diffuses <u>from</u> the blood <u>into</u> the tissue cells. At the same time, since <u>respiratory</u> carbon dioxide is produced as a <u>waste</u> product of cellular metabolism, the carbon dioxide tension within the tissue cells is <u>high</u>. Arterial blood has a <u>low</u> carbon dioxide tension, and the gas therefore diffuses from the <u>tissue</u> cells into the <u>blood</u>.

Test Items

A. 1.b, 2.a, 3.a, 4.a, 5.d, 6.c, 7.a, 8.c, 9.d, 10.d, 11.b, 12.a, 13.d, 14.d, 15.d, 16.d, 17.a, 18.a, 19.a, 20.b.

B. 1.e, 2.b, 3.g, 4.k, 5.d, 6.h, 7.l, 8.m, 9.a, 10.i, 11.n, 12.c, 13.o, 14.j, 15.f.
     1.c, 2.e, 3.g, 4.l, 5.j, 6.o, 7.f, 8.a, 9.n, 10.i, 11.m, 12.h, 13.d, 14.b, 15.k.

C. 1.F, 2.T, 3.T, 4.F, 5.F, 6.T, 7.F, 8.T, 9.T, 10.F, 11.F, 12.F, 13.T, 14.T, 15.T, 16.F, 17.T, 18.F, 19.T, 20.F.

The Respiratory System

Across

1 flap-like structure that covers the opening to the breathing tubes

6 a control center in the pons

8 a branch of the trachea

9 removal of fluid

11 first portion of the breathing tubes—the voice box

13 tenth cranial nerve

14 lymph gland located in the oropharynx

15 membranes formed in the fetal lung

17 difficult breathing

18 the back of the mouth and nose

19 active intake of air

20 space within the chest cavity outside of the lungs

Down

2 pertaining to the diaphram

3 pertaining to the tongue

4 space within the lung alveolar sacs

5 movement of air out of the lungs

7 a small branch of the bronchus

10 receiving chamber of the alveolar cluster

12 air sacs in the lungs

16 dilation and deterioration of the alveloar sacs

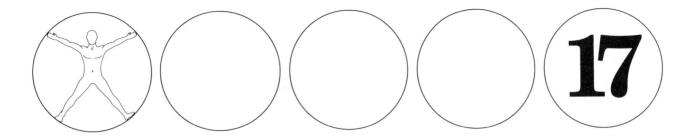

The Alimentary Tract

I. CHAPTER SYNOPSIS

Throughout the text the synthesis of molecules, contraction of muscles, transport of molecules or ions across cell membranes, and maintenance of body temperature are discussed. For these activities, the body depends on its ability to extract and use the chemical potential energy locked within the structure of chemical bonds. Without sufficient energy, the cells die. This chapter deals with the concept of energy, the mechanisms of trapping chemical energy in a form that can be used by cells (ATP synthesis), the properties of enzymes, which catalyze the many chemical reactions in cells, and the general metabolic pathways for the breakdown, synthesis, and interconversion of carbohydrates, fats, and proteins. The student is introduced to metabolism and nutrition.

The digestive process is considered by region, in which the anatomy, physiology, physical processes, and chemical processes of each region are discussed together. There is also emphasis on the nervous and hormonal control of digestion.

II. OBJECTIVES

After reading the chapter, the student should be able to:

- Identify the anatomy of the alimentary tract.
- Explain how saliva is produced and its effect on carbohydrates.

- Describe peristalsis and its control mechanism.

- Differentiate between the cardiac and pyloric stomachs in reference to position and function.

- Define a peptic ulcer and identify where it is found.

- Identify the sites and formation of bile.

- Describe the role of bile in digestion.

- Identify the function of hormones in digestion and give examples.

- Explain how the end products of digestion are absorbed by villi.

- Explain how each of the three foodstuffs are metabolized for energy, synthesis, and storage.

III. IMPORTANT TERMS

Using your textbook, define the following terms:

ampulla (am-pul'-ah) _____

_____

anal (ain'-ul) _____

_____

appendix (ah-pen'-diks) _____

_____

bile (bile) _____

_____

bolus (bo'-lus) _____

_____

carbohydrate (kahr-bo-hi'-drate) _____

_____

defecation (def-eh-kay'-shun) _____

_____

diarrhea (di-ah-ree'-ah) _____

_____

diverticulum (di-vur-tik'-yoo-lum) _____

_____

enzyme (en'-zime) _____

_____

fat (fat) _____

_____

jaundice (jawn'-dis) _____

_____

parietal (pah-ri'-et-ul) _____

_____

protein (pro'-teen) _____

_____

pyloric (pi-lor'-ik) _____

_____

rugae (roo'-guy) _____

_____

sinusoid (si'-nyah-soid) _____

_____

sphincter (sfink'-tur) _____

_____

substrate (sub'-strate) _____

_____

ulcer (ul'-sur) _____

_____

varices (var'-i-seez) _____

_____

vitamin (vite'-ah-min) _____

_____

IV. EXERCISES

Complete the following exercises in the order given. A precise set of terms and diagrams has been chosen to describe the alimentary tract.

Exercise 17.1

Labeling.   Write the name of the structure in the space provided. Color the organs differently.

Key:

anal	liver
appendix	mouth
ascending	parotid
cecum	pharynx
descending	pyloric
diaphragm	rectum
duodenum	sigmoid
esophagus	spleen
gallbladder	stomach
illium	transverse
jejunum	valve

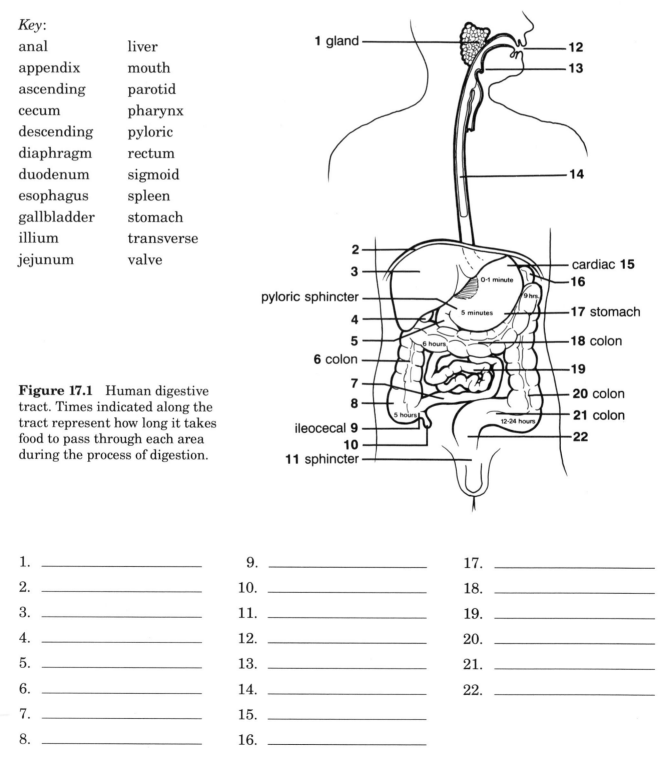

Figure 17.1   Human digestive tract. Times indicated along the tract represent how long it takes food to pass through each area during the process of digestion.

1. _____   9. _____   17. _____

2. _____   10. _____   18. _____

3. _____   11. _____   19. _____

4. _____   12. _____   20. _____

5. _____   13. _____   21. _____

6. _____   14. _____   22. _____

7. _____   15. _____

8. _____   16. _____

Exercise 17.2

Labeling.   Write the name of the structure in the space provided. Color the various layers of the stomach differently.

Key:

cardiac

chief

curvature

duodenum

esophagus

furrow

gastric

mucous

muscle

parietal

pyloric

rugae

smooth

sphincter

submucosa

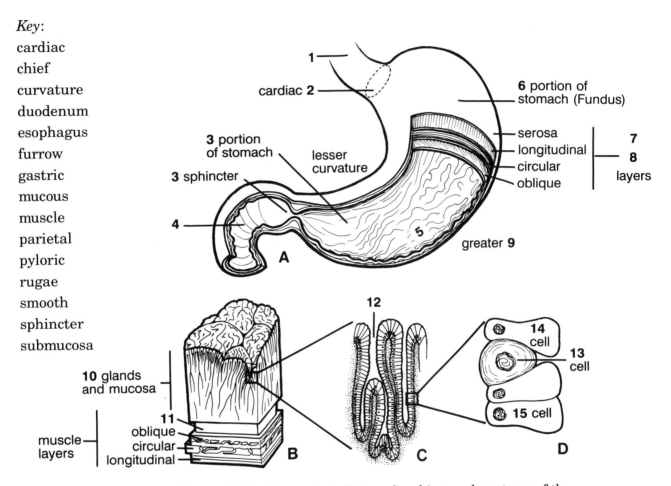

Figure 17.2   Stomach. **A**. External and internal anatomy of the stomach, showing the layers of muscle, the rugae, and the pyloric valve. **B**. Schematic representation of the gastric mucosa, showing all the layers of the stomach. **C**. Gastric glands from the greater curvature of the stomach. **D**. Detail of the gastric glands.

1. _____   9. _____

2. _____   10. _____

3. _____   11. _____

4. _____   12. _____

5. _____   13. _____

6. _____   14. _____

7. _____   15. _____

8. _____

Exercise 17.3

Labeling. Write the name of the structure in the space provided. Color each organ differently.

Key:

ampulla	duodenum	pancreas
bile	gallbladder	pancreatic
cells	hepatic	sinusoids
cystic	Kupffer	
duct	liver	

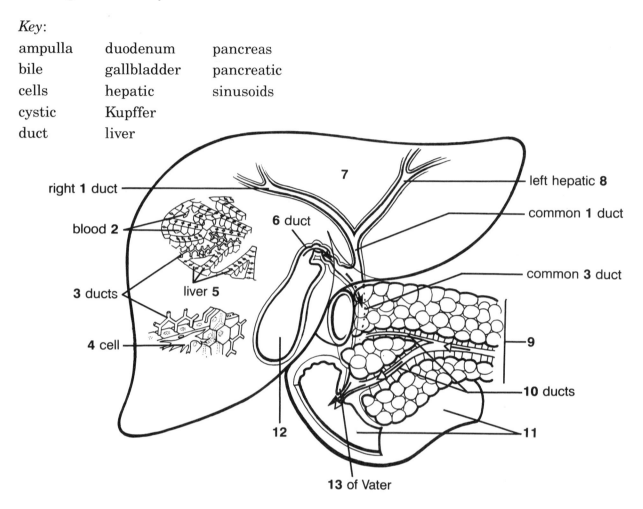

Figure 17.3 Liver and its interrelationship with the gallbladder, pancreas, and duodenum. A section has been removed from the liver and the area enlarged to show the arrangement of liver cells, bile ducts, Kupffer cells, and blood sinusoids to one another. Arrows indicate the direction of flow of bile from the gallbladder and liver and of digestive juices from the pancreas into the duodenum.

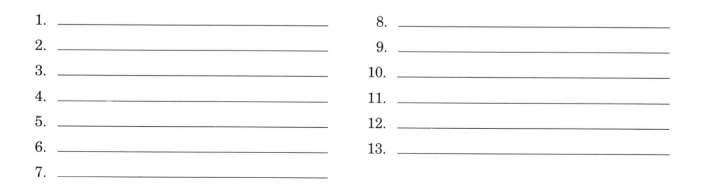

1. _____   8. _____

2. _____   9. _____

3. _____   10. _____

4. _____   11. _____

5. _____   12. _____

6. _____   13. _____

7. _____

Exercise 17.4

Labeling.   Write the name of the hormone or organ in the space provided.
Color the organs in reference to Figures 17.2 and 17.3.

Key:

cholecystokinin	gastrin	secretin
enterogastrone	liver	stomach
enterokinin	pancreas	villikinin
gallbladder	pancreozymine	

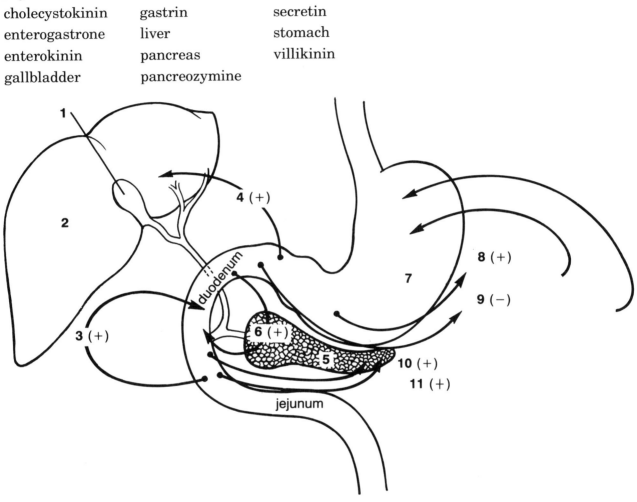

Figure 17.4   Hormonal control of digestion. The (+) sign indicates
stimulation; the (−) sign indicates inhibition.

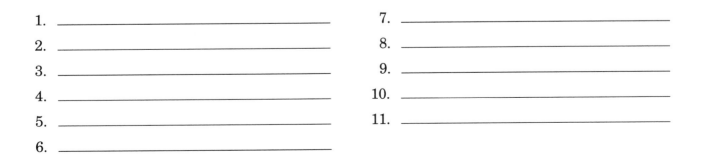

1. _____   7. _____

2. _____   8. _____

3. _____   9. _____

4. _____   10. _____

5. _____   11. _____

6. _____

V. TEST ITEMS

A. *Multiple Choice.*   There is only one answer that is either correct or most appropriate. Circle the answer that corresponds to the question.

1. Inability to catalyze certain chemical reactions in the brain cells might indicate a malfunction of
    a. disaccharides.   c. polysaccharides.
    b. fats.   d. enzymes.

2. If you have just digested fat molecules, you would expect to find increased amounts of
    a. glucose.   c. fatty acids and glycerol.
    b. amino acids.   d. nucleic acids.

3. For your body to manufacture antibodies, certain hormones, and hemoglobin, you would have to include which of the following in your diet?
    a. fats   c. sucrose
    b. proteins   d. polysaccharides

4. If two amino acids are joined together, the resulting molecule is
    a. a protein.   d. a dipeptide.
    b. a disaccharide.   e. an enzyme.
    c. a fatty acid.

5. To remove the pancreas, a surgeon would have to enter which cavity?
    a. pelvic   c. abdominal
    b. thoracic   d. vertebral

6. One of the glands associated with the digestive system is the
    a. spleen.   c. thymus.
    b. liver.   d. hypophysis.

7. In the carbonic acid-bicarbonate buffer system, a decrease in pH is prevented by
    a. sodium bicarbonate.
    b. carbonic acid.
    c. hydrochloric acid.
    d. sodium hydroxide.

8. An obstruction in the ampulla of Vater would affect your ability to transport
    a. bile and pancreatic juice.
    b. gastric juice.
    c. salivary amylase.
    d. succus entericus.

9. Between meals the glucose level of the blood is maintained by
    a. insulin.   c. lipogenesis.
    b. glycogenolysis.   d. glycogenesis.

10. Starvation, low carbohydrate diets, and metabolic abnormalities are all factors that contribute to
    a. ketosis.   c. oxygen debt.
    b. alkalosis.   d. muscle fatigue.

11. The synthesis of glycogen molecules for storage in the liver and skeletal muscles is referred to as
    a. glycogenolysis.
    b. glyconeogenesis.
    c. beta oxidation.
    d. glycogenesis.

12. Blood passing through the sinusoids of a lobule of the liver is brought to the liver by the
    a. central vein and hepatic vein.
    b. hepatic vein and portal vein.
    c. hepatic artery and portal vein.
    d. hepatic artery and hepatic vein.

13. An inability of the body to synthesize cholecystokinin would hamper or inhibit
    a. the secretion of pancreatic juice.
    b. the production of saliva.
    c. the emulsification of fats.
    d. mass peristalsis of the colon.

14. The replacement of hepatic cells by fibrous connective tissue and frequently adipose tissue is called
    a. jaundice.
    b. hepatitis.
    c. peritonitis.
    d. cirrhosis.

15. In which organ is the chemical digestion of carbohydrates initiated?
    a. stomach
    b. small intestine
    c. mouth
    d. large intestine

16. Inflammation of the periodontal membrane and adjacent gingivae is referred to as
    a. peritonitis.
    b. pyorrhea.
    c. mumps.
    d. pancreatitis.

17. To free the small intestine from the posterior abdominal wall, which of the following would have to be cut?
    a. frenulum
    b. mesentery
    c. lesser omentum
    d. falciform ligament

18. Insulin is secreted by the
    a. exocrine portion of the pancreas.
    b. beta cells of the pancreas.
    c. ampulla of Vater.
    d. liver.

19. Trypsinogen is produced by which of the following?
    a. liver
    b. stomach
    c. pancreas
    d. duodenum

20. An x-ray examination of the gallbladder is called a
    a. barium enema.
    b. GI series.
    c. cholecystogram.
    d. gastric analysis.

B. *Matching Questions.* Each of the phrases in COLUMN B refers to a word or phrase in COLUMN A. Insert the letter of the word or phrase from COLUMN B that best describes it. Some words or phrases may be used more than once or not at all.

	Column A		*Column B*
1. ___	calculus	a.	expulsion of the stomach contents through the mouth by reverse peristalsis
2. ___	mumps		
3. ___	cholecystitis	b.	excessive amount of gas in the stomach or intestines
4. ___	diarrhea	c.	inflammation of the gallbladder
5. ___	diverticulosis	d.	infrequent or difficult defecation
6. ___	flatus	e.	a stone in an organ
7. ___	heartburn	f.	inflammation of the colon and rectum
8. ___	constipation	g.	burning sensation in the region of the esophagus and stomach
9. ___	hepatitis		
10. ___	hernia	h.	painful inflammation and enlargement of the salivary glands, particularly the parotids
11. ___	vomiting	i.	frequent defecation of liquid feces
12. ___	pancreatitis	j.	protrusion of an organ or part of an organ through a membrane or wall of a cavity
13. ___	colitis		
14. ___	colostomy	k.	inflammation of the liver
		l.	inflammation of the pancreas
		m.	abnormal sacs or outpockets of the intestinal mucosa into the muscularis
		n.	cutting the colon in half and bringing the upper, lower, or both halves through the abdominal wall to the exterior

C. *True-False.* Place a *T* or *F* in the space provided.

___ **1.** Contractions of the stomach macerate food, mix it, and eventually reduce it to a thin liquid called a bolus.

___ **2.** Cells within gastric glands that secrete hydrochloric acid are called zymogenic cells.

___ **3.** Distention of the stomach is permitted by folds of the mucosa called rugae.

___ **4.** Failure of the muscle fibers of the pyloric valve to relax normally is called pyloric stenosis.

___ **5.** The portion of the stomach closest to the esophagus is the cardia.

___ **6.** Proteins are formed by linking together different types of amino acids in a number of different sequential patterns.

___ **7.** Fatty acids that do not have double bonds in their structures are known as polyunsaturated fatty acids.

___ **8.** The salivary glands whose secretions are discharged on either side of the frenulum are the parotids.

___ **9.** Projections of the lamina propria of the tongue that are covered with epithelium are called papillae.

___ **10.** All enzymes are proteins.

___ **11.** Glycogen, a polysaccharide composed of many molecules of glucose, is an example of an important carbohydrate.

___ **12.** Increased peristaltic activity of the small intestine will lead to diarrhea because of a decreased time for absorption of material.

___ **13.** Two types of contractions, peristalsis and segmenting, occur in the small intestine.

___ **14.** Secretion of digestive enzymes by the stomach is brought about exclusively by hormonal stimulation.

___ **15.** Following their absorption, simple sugars, fats, and amino acids are transported to the liver via the hepatic portal vein.

Answer Sheet—Chapter 17

Exercise 17.1

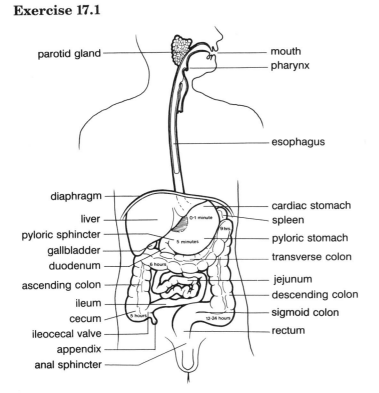

Figure 17.1 Human digestive tract. Times indicated along the tract represent how long it takes food to pass through each area during the process of digestion.

Exercise 17.2

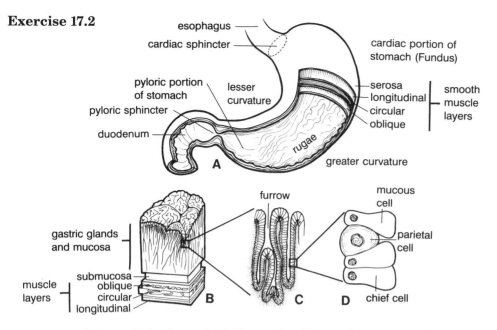

Figure 17.2 Stomach. **A**. External and internal anatomy of the stomach, showing the layers of muscle, the rugae, and the pyloric valve. **B**. Schematic representation of the gastric mucosa, showing all the layers of the stomach. **C**. Gastric glands from the greater curvature of the stomach. **D**. Detail of the gastric glands.

Exercise 17.3

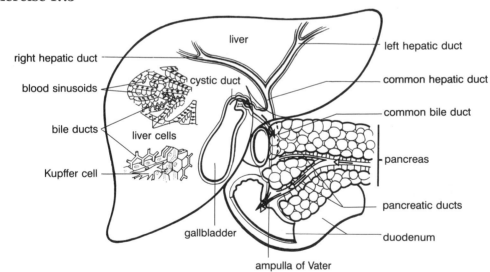

Figure 17.3 Liver and its interrelationship with the gallbladder, pancreas, and duodenum. A section has been removed from the liver and the area enlarged to show the arrangement of liver cells, bile ducts, Kupffer cells, and blood sinusoids to one another. Arrows indicate the direction of flow of bile from the gallbladder and liver and of digestive juices from the pancreas into the duodenum.

Exercise 17.4

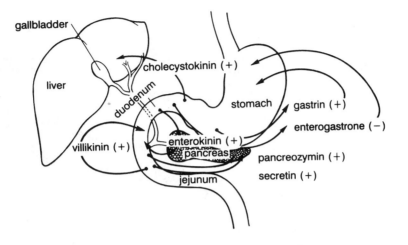

Figure 17.4 Hormonal control of digestion. The (+) sign
indicates stimulation; the (−) sign indicates inhibition.

Test Items

A. 1.d, 2.c, 3.b, 4.d, 5.c, 6.b, 7.a, 8.a, 9.b, 10.a, 11.d, 12.c, 13.c, 14.d, 15.c, 16.b, 17.b, 18.b,
19.c, 20.c.

B. 1.e, 2.h, 3.c, 4.i, 5.m, 6.b, 7.g, 8.d, 9.k, 10.j, 11.a, 12.l, 13.f, 14.n.

C. 1.F, 2.F, 3.T, 4.F, 5.T, 6.T, 7.F, 8.F, 9.T, 10.T, 11.T, 12.T, 13.T, 14.F, 15.F.

Digestion

Across

1 semi-digested food mass from the mouth

3 the last part of the intestine

4 an organic catalyst

6 the lower part of the stomach

8 organic substance needed for metabolism

12 expulsion of fecal material from the anus

14 folds of tissue on the inner wall of the stomach

15 pertaining to the wall or outer part of a cavity

16 any substance acted upon by an enzyme

17 abnormally swollen veins

Down

1 a yellow-green waste product of the liver

2 principal energy food source

5 a circular smooth muscle arrangement

7 an outpouching along the digestive tract

9 yellow color of the skin

10 abnormal liquid discharge from the anus

11 a lesion of mucous membranes

13 storage compounds made up of fatty acids and glycerol

15 nitrogenous building block of the cell

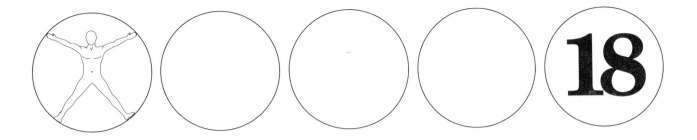

Excretion

I. CHAPTER SYNOPSIS

As seen in Chapter 16, the lungs excrete the major metabolic waste product of the body—carbon dioxide. Other waste products, especially urea, the end product of nitrogen metabolism, are excreted by the kidneys into the urine. However, the excretion of waste products is of secondary importance in relation to the role of the kidneys in the regulation of the salt and water content of the internal environment. The salt and water content of the body depends upon the balance between intake and excretion. The major control point is the regulation of excretion by the kidneys. The amounts of water and salt excreted by the kidneys depend upon the magnitude of the glomerular filtration, tubular reabsorption, and secretion. These basic operations of the kidney are influenced by several hormones whose release is determined by the amounts of salt and water in the blood. Thus, this chapter examines the structure of the kidney and urinary system, basic principles of renal physiology, micturition, and kidney disease.

II. OBJECTIVES

After reading the chapter, the student should be able to:

- Label the anatomy of the urinary system.

- Identify all urinary structures and relate their separate functions.

- Define micturition and identify the factors involved in glomerular filtration rates.

- Define diuresis and explain how it works.

- Explain the role of ADH and aldosterone in renal tubular absorption.

- Identify the active and passive processes occurring during tubular reabsorption.

- Explain how renal failure disturbs the body's homeostasis.

- Define nephritis, glomerulonephritis, pyelonephritis, and urethritis.

- Explain dialysis and identify the two major methods.

III. IMPORTANT TERMS

Using your textbook, define the following terms:

adrenal (ah-dreen'-ul) _____

_____

antidiuretic (ant-eh-die-yoo-ret'-ik) _____

_____

calculi (kal'-kew-lie) _____

_____

calyx (kay'-liks) _____

_____

cystoscope (sis'-tah-skope) _____

_____

dehydration (dee-hi-dray'-shun) _____

_____

diuresis (di-yah-ree'-sis) _____

_____

dysuria (dish-yoor'-ee-ah) _____

_____

excretion (ek-skree'-shun) _____

_____

glomerulus (glah-mer'-yah-lus) _____

_____

hemodialysis (hee-mo-di-al'-ah-sis) _____

_____

hilus (hi'-lus) _____

_____

incontinence (in-kant'-in-ahnts) _____

_____

micturition (mik-chah-rish'-un) _____

_____

nephron (nef'-ron) _____

_____

pyramid (pir'-ah-mid) _____

_____

renal (reen'-ul) _____

_____

retroperitoneal (reh-tro-per-ah-tone-ee'-ul) _____

_____

transitional (trants-ish'-un-ul) _____

_____

tubular (tew'-byah-lur) _____

_____

urea (yoo-ree'-ah) _____

uremia (yoo-ree'-mee-ah) _____

_____

urination (yoor-ah-nay'-shun) _____

_____

IV. EXERCISES

Complete the following exercises in the order given. A precise set of terms and diagrams has been chosen to describe excretion.

Exercise 18.1

Labeling.   Write the name of the structure in the space provided. Color the vessels and organs separately.

Key:
adrenal
aorta
artery
external
iliac
kidney
prostate
renal
ureters
urethra
urinary
vena cava

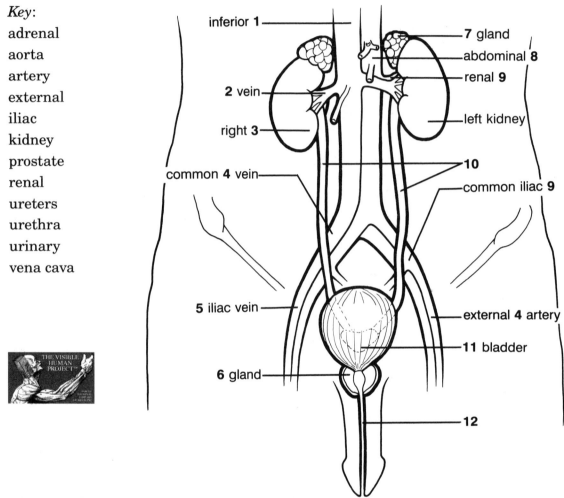

Figure 18.1   Urinary system with blood vessels.

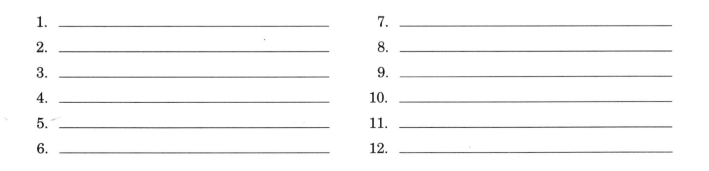

1. _____   7. _____

2. _____   8. _____

3. _____   9. _____

4. _____   10. _____

5. _____   11. _____

6. _____   12. _____

Exercise 18.2

Labeling.   Write the name of the structure in the space provided. Color the three areas of the kidney differently.

Key:

calyx	minor
capsule	opening
cortex	pelvis
fibrous	pyramids
hilus	renal
medulla	ureter

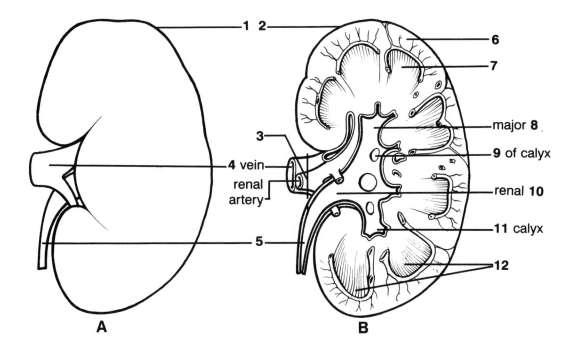

Figure 18.2   External (**A**) and internal (**B**) anatomy of the left kidney.

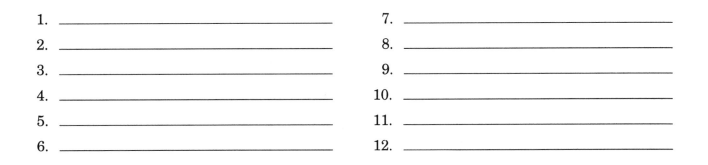

1. _____

2. _____

3. _____

4. _____

5. _____

6. _____

7. _____

8. _____

9. _____

10. _____

11. _____

12. _____

Exercise 18.3

Labeling.   Write the name of the structure in the space provided. Color the vessels differently from the tubules.

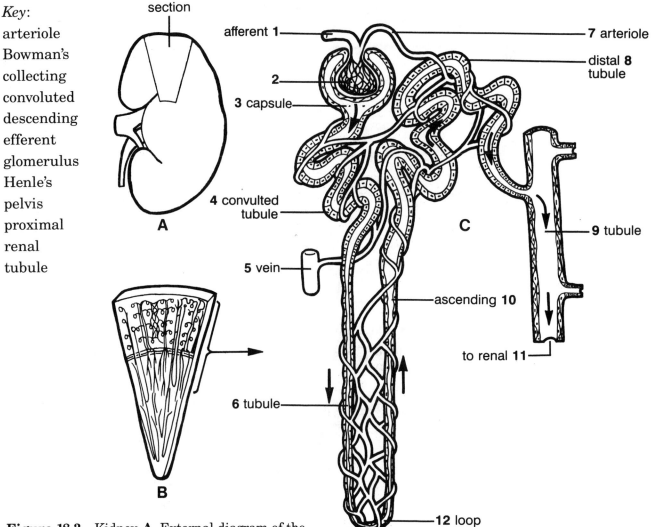

Key:

arteriole
Bowman's
collecting
convoluted
descending
efferent
glomerulus
Henle's
pelvis
proximal
renal
tubule

Figure 18.3   Kidney. **A**. External diagram of the left kidney. **B**. and **C**. A section of the kidney has been enlarged to show the details of a nephron unit. Arrows indicate the direction of flow of urine through the nephron unit.

1. _____

2. _____

3. _____

4. _____

5. _____

6. _____

7. _____

8. _____

9. _____

10. _____

11. _____

12. _____

Exercise 18.4

Labeling. Write the name of the structure in the space provided. Color the organs and tubes differently.

Key:
connective
Cowper's
fibrous
internal
muscle
prostate
smooth
sphincter
transitional
ureter
ureters
urethra

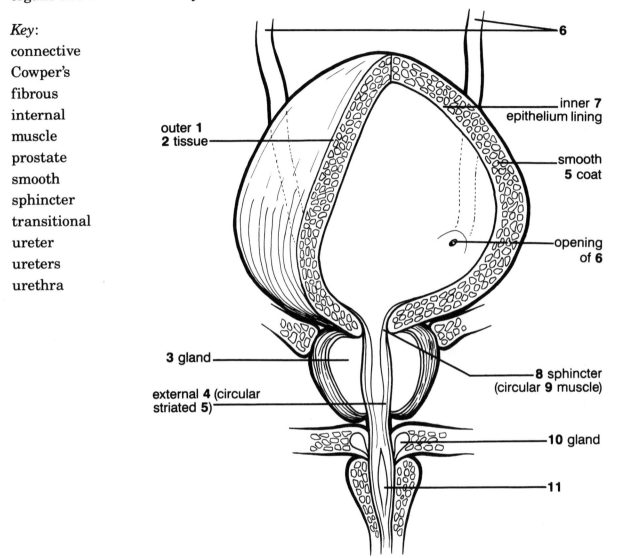

Figure 18.4 Urinary bladder, ureters, and urethra of an adult male.

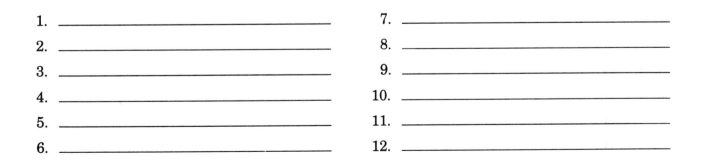

1. _____ 7. _____

2. _____ 8. _____

3. _____ 9. _____

4. _____ 10. _____

5. _____ 11. _____

6. _____ 12. _____

V. TEST ITEMS

A. *Multiple Choice.*   There is only one answer that is either correct or most appropriate. Circle the answer that corresponds to the question.

1. Damage to the renal medulla would interfere first with the functioning of the
   a. Bowman's capsule.
   b. distal convoluted tubule.
   c. collecting ducts.
   d. proximal convoluted tube.

2. The ability to store urine would be affected by which of the following conditions?
   a. pyelitis
   b. nephrosis
   c. cystitis
   d. renal suppression

3. An obstruction in the glomerulus would affect the flow of blood into the
   a. renal artery.
   b. efferent arteriole.
   c. afferent arteriole.
   d. intralobular artery.

4. Fluid output normally is greatest through which of the following?
   a. skin
   b. kidneys
   c. gastrointestinal tract
   d. lungs

5. The rate of filtration
   a. is dependent of the hydrostatic pressure.
   b. increases with increasing size of the pores in the filter.
   c. through the capillary walls is the same in all parts of the body.
   d. all of the above

6. The most abundant cation in extracellular fluid is
   a. Na^+.
   b. K^+.
   c. Cl^-.
   d. HPO_4^-.

7. Urine that leaves the distal convoluted tubule passes through the following structure in which sequence?
   a. collecting duct, hilum, calyx, ureter
   b. collecting duct, calyx, pelvis, ureter
   c. calyx, collecting duct, pelvis, ureter
   d. calyx, hilum, pelvis, ureter

8. In the carbonic acid-sodium bicarbonate buffer system, strong acids are buffered by the
   a. carbonic acid.
   b. salt of the weak acid.
   c. water.
   d. sodium chloride.

9. A high level of uric acid in the blood that may crystallize in the kidneys and joints results in a disorder called
   a. pyuria.
   b. gout.
   c. pyelonephritis.
   d. anuria.

10. As the amount of vasopressin-ADH in the blood increases,
    a. urine is made more acidic.
    b. more urine is formed.
    c. tubular reabsorption of water increases.
    d. diuresis is accelerated.

11. Under normal circumstances, filtrate passing from the glomerulus into Bowman's capsule is free of
    a. water.
    b. glucose.
    c. proteins.
    d. chloride.

12. The kidney control of pH is based upon the excretion of the ammonium ion in exchange for the reabsorption of
    a. potassium.
    b. chloride.
    c. hydrogen.
    d. sodium.

13. Hypertension and inflammation are two conditions that may lead to
    a. dehydration.
    b. edema.
    c. excessive water loss.
    d. alkalosis.

14. Conditions such as diabetes mellitus, starvation, and low carbohydrate diets are closely linked to
    a. ketosis.
    b. pyuria.
    c. hematuria.
    d. calculi.

15. Incontinence is distinguished from retention in that the former is a(n)
    a. failure to void urine.
    b. inability of the kidneys to make urine.
    c. lack of voluntary control over micturition.
    d. voluntarily controlled by the cerebral cortex.

16. The primary action of ADH is to
    a. cause active reabsorption of sodium.
    b. cause active reabsorption of water.
    c. increase the permeability of the ascending limb of the loop of Henle.
    d. increase the permeability of the collecting tube.

17. Osmoreceptors in the hypothalamus control
    a. release of ADH.
    b. release of aldosterone.
    c. electrolyte reabsorption.
    d. glucose reabsorption.

18. The kidney contributes to the maintenance of acid-base balance primarily by
    a. secreting urea.
    b. excreting H^+ in exchange for Na^+.
    c. excreting Na^+ in exchange for H^+.
    d. excreting HCO_3^-.

19. Another means by which the kidney controls the pH of the blood is by
    a. excreting CO_2.
    b. excreting the buffers that transport acid substances in the blood.
    c. excreting Na^+.
    d. secreting ammonia.

20. Reabsorption occurs in the
    a. glomerulus.
    b. afferent arteriole.
    c. capillary net about the tubule.
    d. renal vein.

B. *Matching Questions.* Each of the phrases in COLUMN B refers to a word or phrase in COLUMN A. Insert the letter of the word or phrase from COLUMN B that best describes it. Some of the words or phrases may be used more than once or not at all.

Column A	*Column B*
1. ___ glomerulus	**a.** the deep fissure on the kidney
2. ___ renal pyramids	**b.** cone-shaped masses in the medulla
3. ___ micturition	**c.** epithelial cells
4. ___ aldosterone	**d.** a polysaccharide
5. ___ hypocalcemic tetany	**e.** urination
	f. ADH
6. ___ ureter	**g.** stimulates sodium reabsorption
7. ___ HCl	**h.** drains urine from the kidneys
8. ___ lactic	**i.** drains urine from the bladder
9. ___ hemoglobin	**j.** a disease caused by low calcium
10. ___ urethra	**k.** plays a role in calcium metabolism
11. ___ insulin	**l.** a strong acid
12. ___ hilum	**m.** the vascular component of the nephron
13. ___ vasopressin	**n.** a weak acid
14. ___ vitamin D	**o.** a buffer
15. ___ podocytes	

Column A	*Column B*
1. ___ respiratory alkalosis	**a.** abnormal decrease in pH due to a decrease in the rate of respiration
2. ___ buffer system	**b.** a chemical substance that dissociates ions
3. ___ edema	**c.** a weak acid and salt of that acid that prevents drastic changes in pH
4. ___ metabolic acidosis	
5. ___ electrolyte	**d.** abnormal decrease in pH due to the buildup of metabolic acids in the blood and/or loss of bicarbonate
6. ___ extracellular fluid	
7. ___ respiratory acidosis	**e.** abnormal increase in pH due to an increase in the minute volume of respiration
	f. a larger than normal volume of interstitial fluid produces swelling of the tissue
8. ___ intracellular fluid	
9. ___ metabolic alkalosis	**g.** body fluid found inside cells
	h. abnormal increase in pH due to a loss of acid by the body or excessive intake of alkaline substances
	i. fluid outside of body cells such as plasma and interstitial fluid

C. *True-False.*   Place a *T* or *F* in the space provided.

____ **1.** In the normal adult, the glomerular filtration rate is about 500 mL/minute.

____ **2.** The term renal suppression applies to a condition in which glomerular blood hydrostatic pressure falls to about 50 mmHg.

____ **3.** Because the afferent arteriole is smaller in diameter than the efferent arteriole, it offers more resistance to the outflow of blood from the glomerulus.

____ **4.** The three steps involved in urine formation are filtration, reabsorption, and secretion.

____ **5.** A renal corpuscle is a Bowman's capsule together with its glomerulus.

____ **6.** The part of a nephron that passes the filtrate into the collecting duct is the proximal convoluted tube.

____ **7.** The triangular-shaped structures inside the medulla of a kidney are called the calyces.

____ **8.** Inflammation of the kidney pelvis and its calyces is referred to as pyelitis.

____ **9.** Urine always has an acid reaction.

____ **10.** Angiotensin II stimulates aldosterone secretion and causes vasoconstriction and elevation of blood pressure.

____ **11.** Glomeruli are small arteries present in the kidney.

____ **12.** The volume of urine secreted is regulated mainly by mechanisms that control the glomerular filtration rate.

____ **13.** A decrease in glomerular blood pressure tends to decrease glomerular filtration rate.

____ **14.** An increase in the hydrostatic pressure in the Bowman's capsule tends to increase the glomerular filtration rate.

____ **15.** A decrease in blood protein concentration tends to decrease the glomerular filtration rate.

____ **16.** Osmoreceptors lie in the anterior portion of the hypothalamus.

____ **17.** The functioning renal unit of the kidney is called the nephron.

____ **18.** Glomerular capillary pressure is higher than the blood pressure in capillaries in other body tissues.

____ **19.** The kidneys lie behind the peritoneum, that is, retroperitoneally.

____ **20.** The bases of the pyramids rest on the renal medulla.

Answer Sheet—Chapter 18

Exercise 18.1

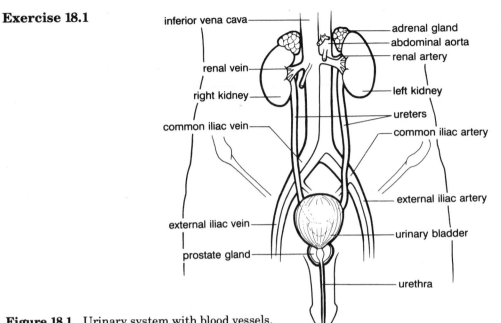

inferior vena cava

renal vein

right kidney

common iliac vein

external iliac vein

prostate gland

adrenal gland
abdominal aorta
renal artery

left kidney

ureters

common iliac artery

external iliac artery

urinary bladder

urethra

Figure 18.1 Urinary system with blood vessels.

Exercise 18.2

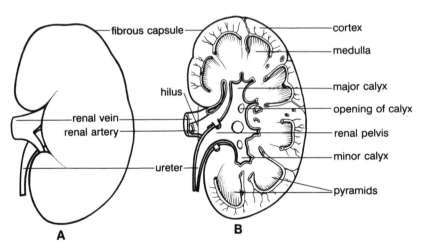

fibrous capsule

hilus

renal vein
renal artery

ureter

cortex

medulla

major calyx

opening of calyx

renal pelvis

minor calyx

pyramids

A

B

Figure 18.2 External (**A**) and internal (**B**) anatomy of the left kidney.

Welcome to the Visible Human

The Visible Human Project is the most complete computerized database of the human body ever assembled. Called "the greatest contribution to anatomy since Vesalius's 1543 publication of De Humani Corpuris Fabrica," the first collection of faithfully rendered drawings of human anatomy, the images are the seeds of a growing medical revolution. Those already familiar with anatomy can follow a structure through different views of the same body, seeing how an organ or muscle changes in space. For the lay audience it is a window into the unknown, a way to visualize the body's mysterious machinery, to see what the professionals see.

The Visible Human Project was conceived by the National Library of Medicine (NLM) in 1988 after bringing together eight medical centers working in the area of 3-D anatomical visualization. Collectively, these centers decided that the NLM should create an image database of human anatomy. The goal was to obtain a perfect, uniform set of photographs and scans from one optimum specimen. Over 100 medical schools vied for the job, and a proposal submitted by the University of Colorado, the home of Dr. Victor Spitzer, won the contract.

The first model, presented here, is a middle aged, Caucasian male who, aside from a few minor imperfections (no left testicle, appendix, or number 14 tooth), was well suited for documentation. To maximize clinical and educational value the virtual subject is documented in several different formats commonly used by radiologists and other physicians: traditional x-rays and computerized tomography scans to visualize bone; magnetic resonance imaging for soft tissue; and three types of color photographs. These make up the multispectral database of images. *The Human Anatomy and Physiology Coloring Workbook and Study Guide, Second Edition* contains fifteen pages of newly generated images compiled from the Visible Human Database. These images are today's version of a dissection in a public amphitheater, but ours is a global amphitheater able to serve an unlimited audience.

The order of images is as follows:

1. Skeleton	6. Brain	11. Heart
2. Total Knee	7. Spinal Cord	12. Circulatory System
3. Knee Joint	8. Endocrine System	13. Respiratory System
4. Muscle	9. Eye	14. Digestive System
5. Central Nervous System	10. Ear	15. Urinary System

For more information on Jones and Bartlett's Visible Human text and CD-ROM products visit Jones and Bartlett's Visible Human World Wide Web site (www.jbpub.com/visiblehuman/). From there, link to the National Library of Medicine's Visible Human Project home page for more information on the Visible Human Project and instructions for obtaining and viewing Visible Human data.

The real future of medical education lies not in the Visible Human data itself but rather in the manipulation, distortion and modification of the data to produce whole populations of virtual humans of every age, race and pathology. With *that* information the possibilities are almost endless.

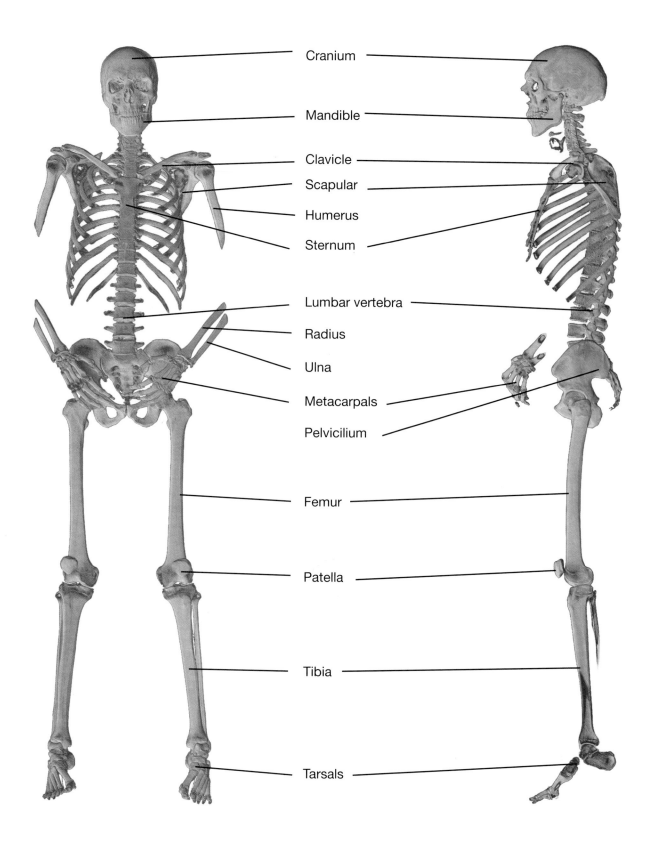

Cranium

Mandible

Clavicle

Scapular

Humerus

Sternum

Lumbar vertebra

Radius

Ulna

Metacarpals

Pelvicilium

Femur

Patella

Tibia

Tarsals

Plate 1 / Skeleton

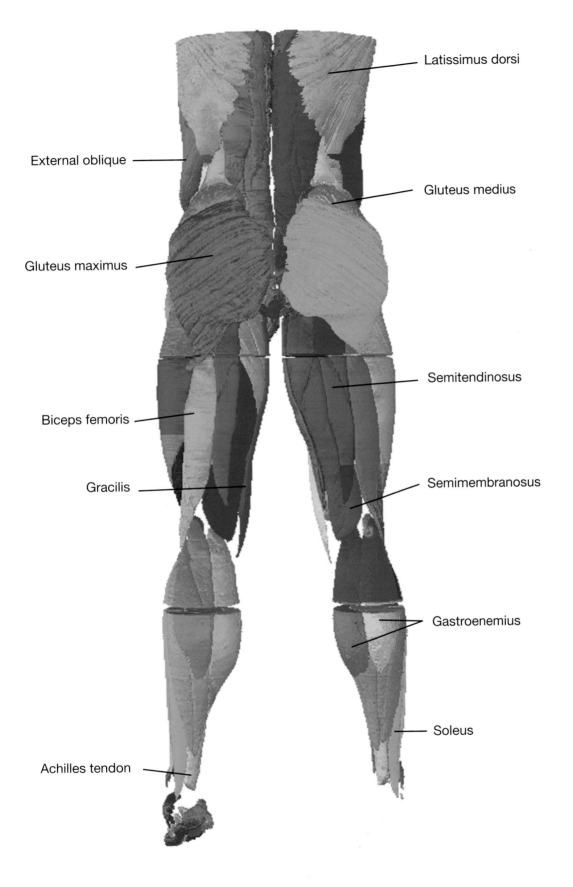

Latissimus dorsi

External oblique

Gluteus medius

Gluteus maximus

Semitendinosus

Biceps femoris

Gracilis

Semimembranosus

Gastroenemius

Soleus

Achilles tendon

Plate 4 / Muscle

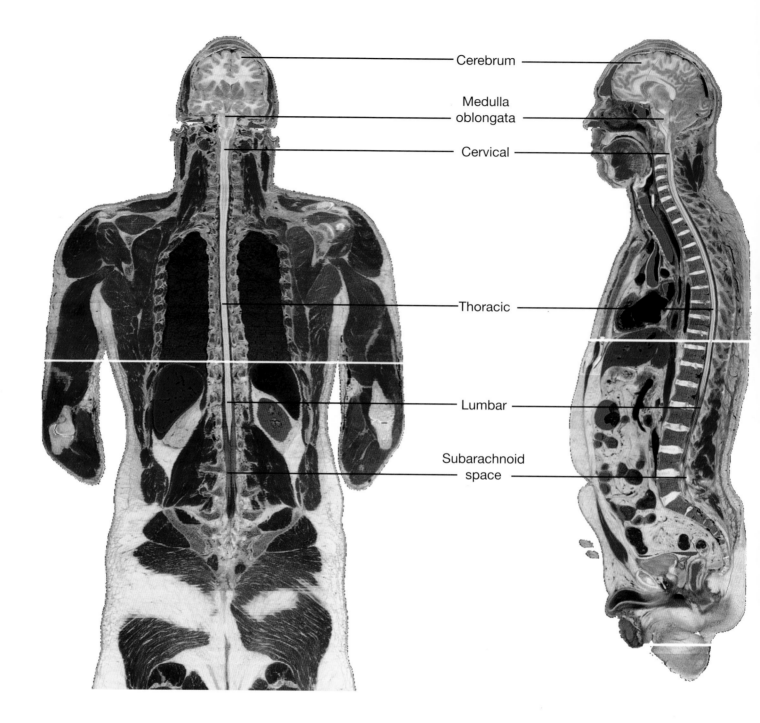

Cerebrum

Medulla
oblongata

Cervical

Thoracic

Lumbar

Subarachnoid
space

Plate 5 / Central Nervous System

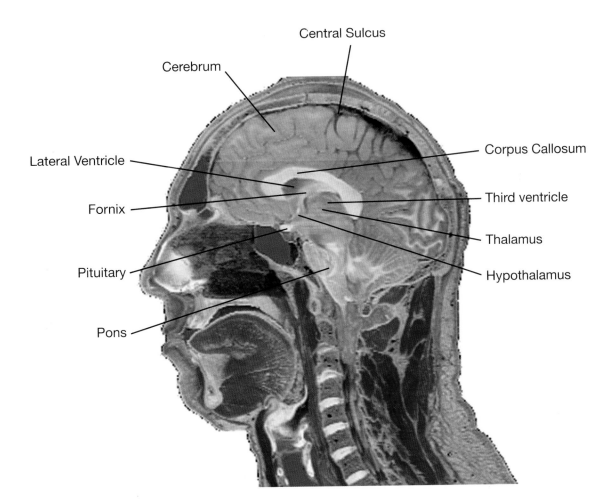

Central Sulcus

Cerebrum

Lateral Ventricle

Fornix

Pituitary

Pons

Corpus Callosum

Third ventricle

Thalamus

Hypothalamus

Plate 6 / Brain

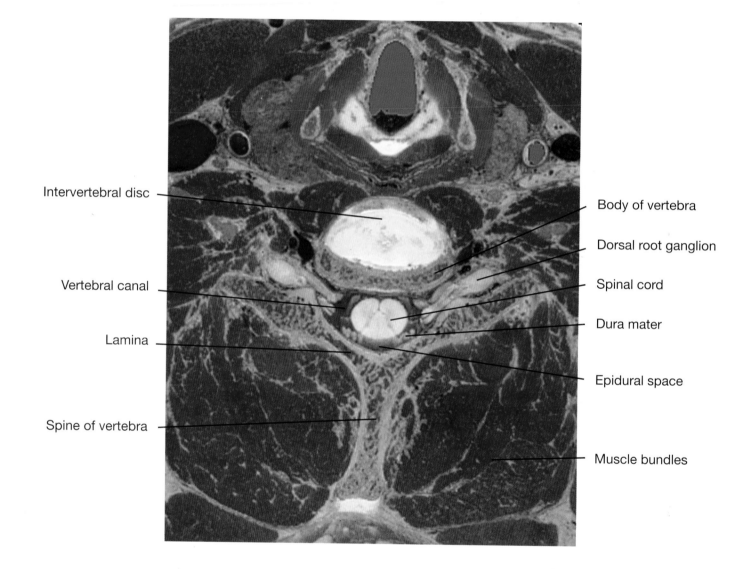

Intervertebral disc

Vertebral canal

Lamina

Spine of vertebra

Body of vertebra

Dorsal root ganglion

Spinal cord

Dura mater

Epidural space

Muscle bundles

Plate 7 / Spinal cord

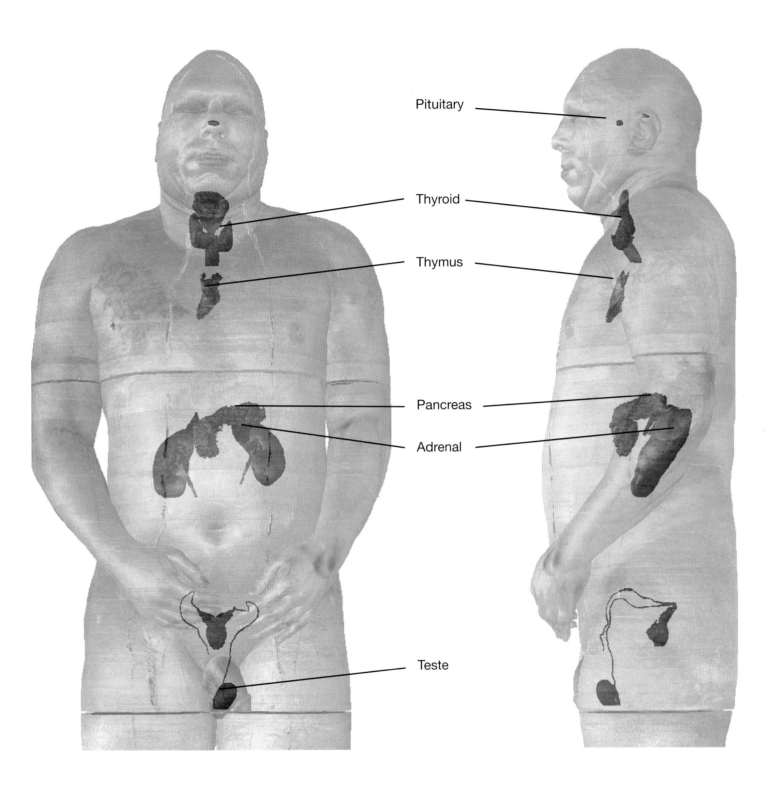

Pituitary

Thyroid

Thymus

Pancreas

Adrenal

Teste

Plate 8 / Endocrine System

Nasal bone

Maxillary bone

Nasal cavity

Hypothalamus

Whitematter

Gray matter

Temporal lobe

Cerebral cortex

Occipital lobe

Occipital bone

Right eyeball

Optic nerve

Optic chiasma

Optic tract

Third ventricle

Lateral ventricle

Cerebellar cortex

Optic radiations

Longitudinal fissure

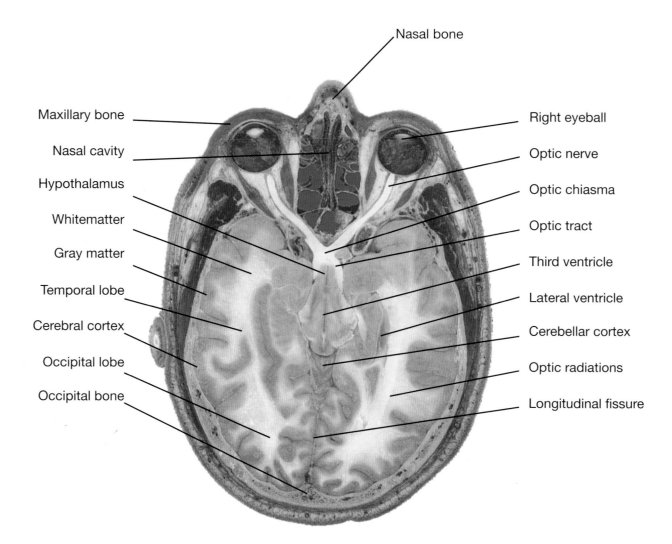

Plate 9 / Eye

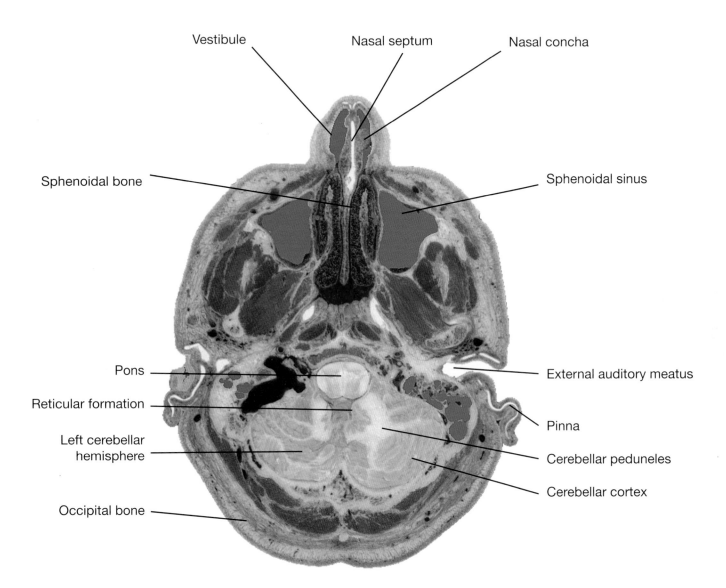

Vestibule

Nasal septum

Nasal concha

Sphenoidal bone

Sphenoidal sinus

Pons

External auditory meatus

Reticular formation

Pinna

Left cerebellar
hemisphere

Cerebellar peduneles

Cerebellar cortex

Occipital bone

Plate 10 / Ear

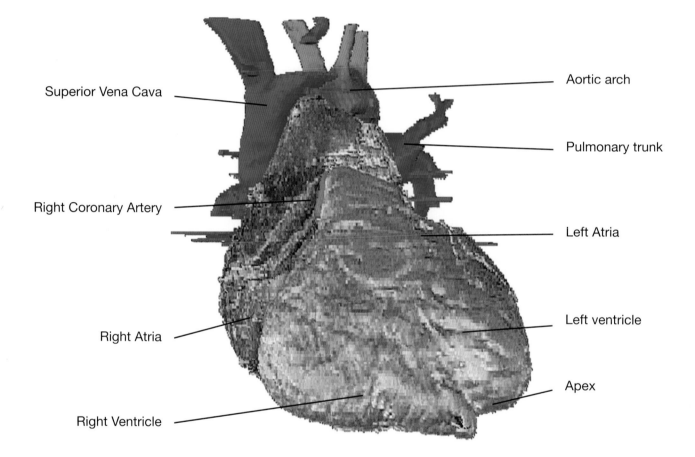

Superior Vena Cava

Aortic arch

Pulmonary trunk

Right Coronary Artery

Left Atria

Right Atria

Left ventricle

Apex

Right Ventricle

Plate 11 / Heart

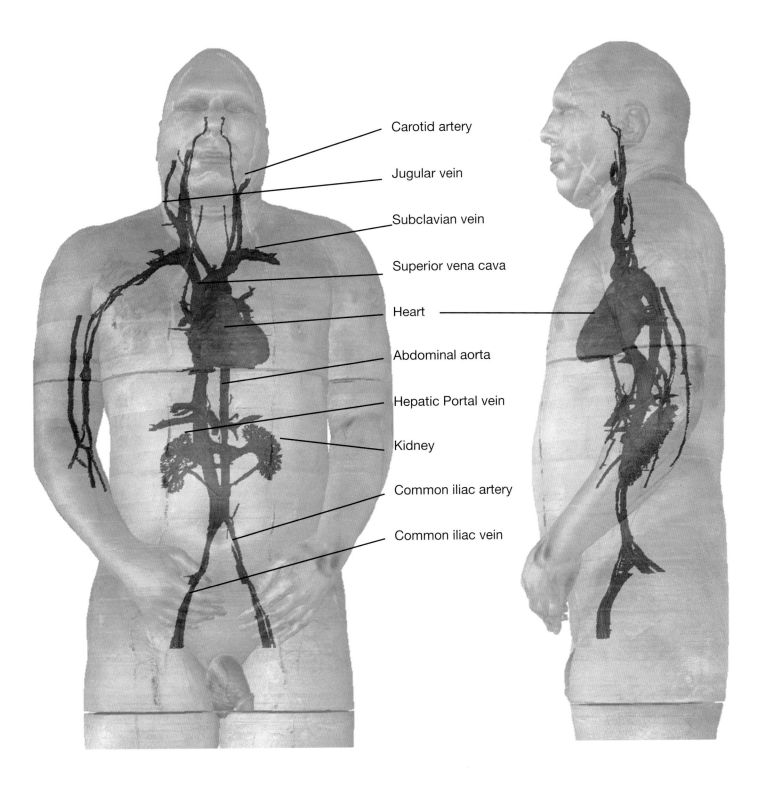

Carotid artery

Jugular vein

Subclavian vein

Superior vena cava

Heart

Abdominal aorta

Hepatic Portal vein

Kidney

Common iliac artery

Common iliac vein

Plate 12 / Circulation

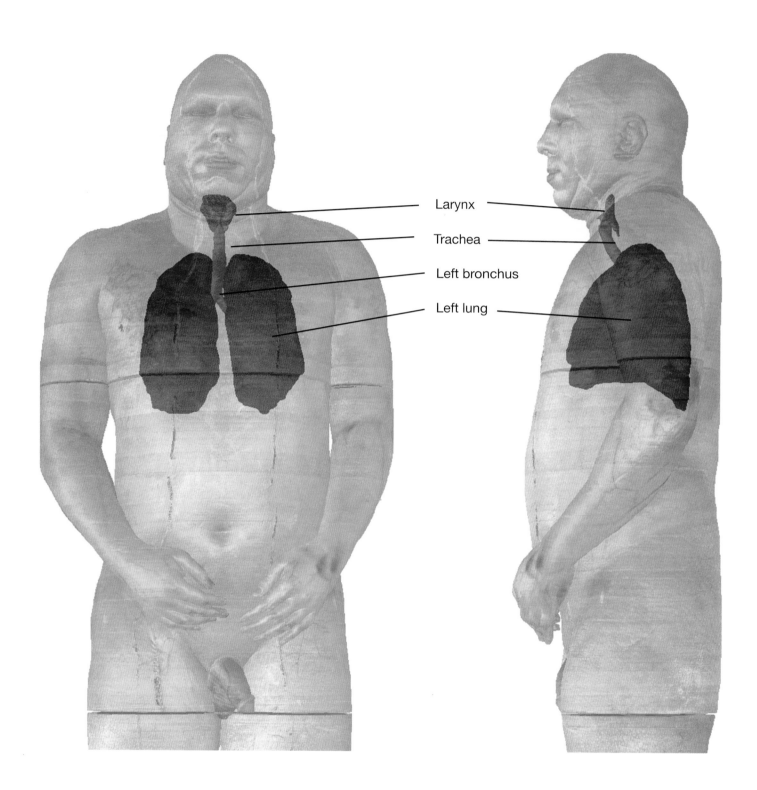

Larynx

Trachea

Left bronchus

Left lung

Plate 13 / Resp.

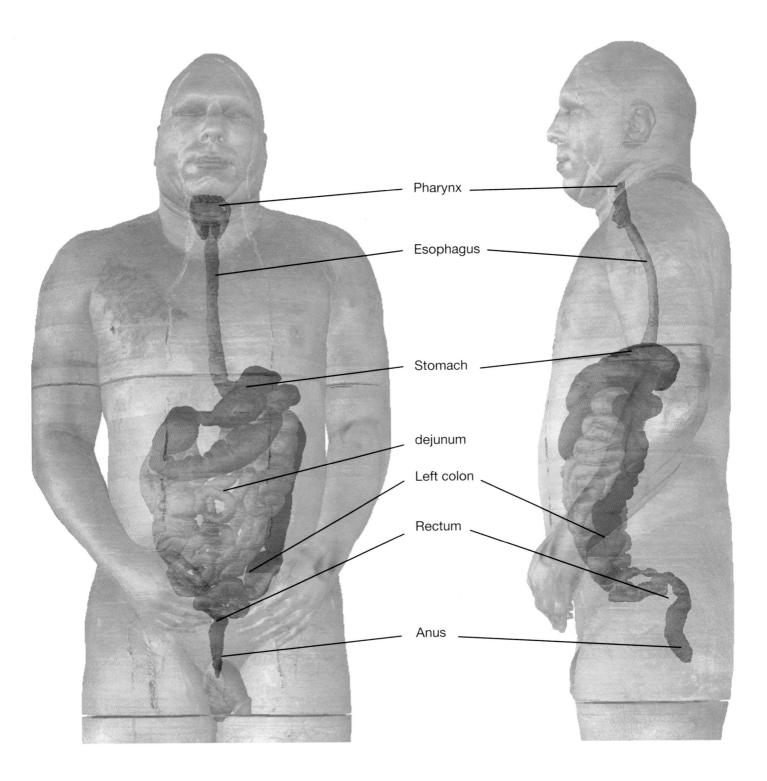

Pharynx

Esophagus

Stomach

dejunum

Left colon

Rectum

Anus

Plate 14 / Digest

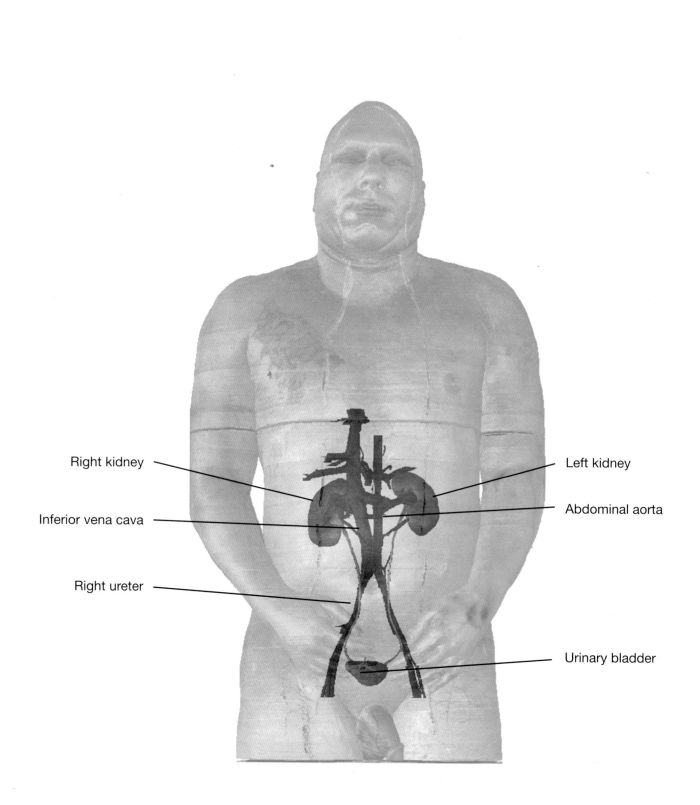

Right kidney

Left kidney

Inferior vena cava

Abdominal aorta

Right ureter

Urinary bladder

Plate 15 / Urinary

Exercise 18.3

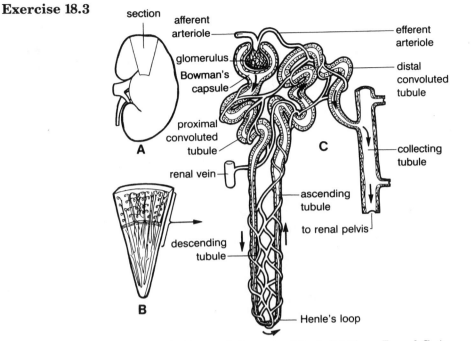

Figure 18.3 Kidney. **A**. External diagram of the left kidney. **B**. and **C**. A section of the kidney has been enlarged to show the details of a nephron unit. Arrows indicate the direction of flow of urine through the nephron unit.

Exercise 18.4

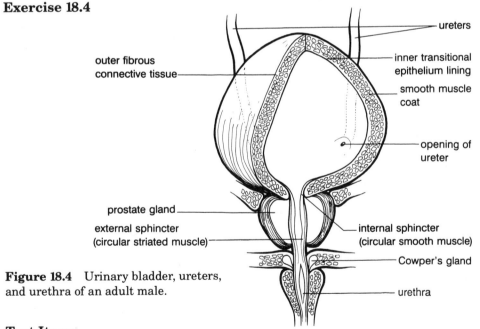

Figure 18.4 Urinary bladder, ureters, and urethra of an adult male.

Test Items

A. 1.c, 2.c, 3.b, 4.b, 5.a, 6.a, 7.b, 8.b, 9.b, 10.c, 11.c, 12.d, 13.b, 14.a, 15.c, 16.d, 17.a, 18.b, 19.d, 20.c.

B. 1.m, 2.b, 3.e, 4.g, 5.j, 6.h, 7.l, 8.n, 9.o, 10.i, 11.d, 12.a, 13.f, 14.k, 15.c.
1.e, 2.c, 3.f, 4.d, 5.b, 6.i, 7.a, 8.g, 9.h.

C. 1.F, 2.T, 3.F, 4.T, 5.T, 6.F, 7.F, 8.F, 9.F, 10.T, 11.F, 12.F, 13.T, 14.F, 15.F, 16.T, 17.T, 18.T, 19.T, 20.F.

Excretion

Across

2 toxic substances in the blood

3 an organ attached to the kidney

4 pertaining to the kidney

7 inadequate amount of water in the body

10 waste product of protein metabolism

12 removal of substances from the cell

13 stones formed within the body

15 removal of wastes from the blood using a semi-permeable membrane

16 an instrument used to examine the inside of the bladder

Down

1 functional unit of the kidney

3 a chemical that prevents water loss

5 difficulty in releasing urine

6 urine output

8 urination

9 network of capillaries in the kidney that filter blood

11 inability to control the passage of urine or feces

14 a funnel-shaped structure in the kidney

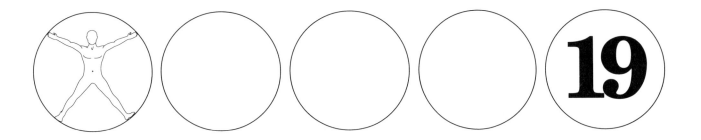

Reproduction

I. CHAPTER SYNOPSIS

Although the reproductive organs of the male and female are not essential to the maintenance of the living state (regulation of the internal environment), they are essential for maintaining the life and continuity of the species. The student is introduced to the anatomy and physiology of the organs of reproduction. Emphasis is placed on the endocrine relations of the male and female systems, particularly those of the menstrual dysfunctions, ovarian cysts, leukorrhea, tumors of the breasts, and cervical cancer.

II. OBJECTIVES

After reading the chapter, the student should be able to:

- Contrast mitosis and meiosis in terms of tissue, location, and number of chromosomes.

- Define spermatogenesis and explain the role of the pituitary gland.

- Identify the structures of the male and female reproductive systems.

- Contrast spermatogenesis and oogenesis.

- List the constituents of semen.

- Name the ovarian and pituitary hormones that regulate menstruation.

- Define myometrium, endometrium, and endometriosis.

- Describe the changes that occur in the uterus during menstruation.

- Define puberty, menopause, and menarche in relation to hormonal production and cessation.

- Describe the glandular structure of the breast and the hormonal influences of lactation.

- Explain the suckling phenomenon.

III. IMPORTANT TERMS

Using your textbook, define the following terms:

abortion (ah-bor'-shun) _____

_____

alveolus (al-vee'-o-lus) _____

_____

ampulla (am-pul'-ah) _____

_____

autosome (awt'-o-som) _____

_____

chromosome (kro'-mah-som) _____

_____

copulation (kop-yah-lay'-shun) _____

_____

corpus (kor'-pus) _____

_____

cryptorchidism (krip-tor'-kah-diz-em) _____

_____

dysmenorrhea (dis-men-ah-ree'-ah) _____

_____

ectopic (ek-top'-ik) _____

_____

ejaculation (eh-jak-yah-lay'-shun) _____

_____

endometriosis (en-do-mee-tree-o'-sis) _____

_____

fertilization (furt-il-ah-zay′-shun) _____

_____

fetus (feet′-us) _____

_____

gamete (gam′-eet) _____

_____

lactation (lak-tay′-shun) _____

_____

meiosis (mi-o′-sis) _____

_____

menopause (men′-ah-pawz) _____

_____

menstruation (ment-strah-way′-shun) _____

_____

mitosis (mi-to′-sis) _____

_____

myometrium (mi-o-mee′-tree-um) _____

_____

ovulation (ahv-yah-lay′-shun) _____

_____

ovum (o′-vum) _____

_____

proliferation (prah-lif-ah-ray′-shun) _____

_____

semen (see′-men) _____

_____

somatic (so-mat′-ik) _____

_____

sperm (spurm) _____

_____

vasectomy (vah-sek′-tah-mee) _____

_____

zygote (zi′-goat) _____

_____

IV. EXERCISES

Complete the following exercises in the order given. A precise set of terms and diagrams has been chosen to describe the reproductive system.

Exercise 19.1

Labeling. Write the term in the space provided. Color the various stages differently.

Key:
fertilization
growth
maturation
meiosis
oocyte
oogenesis
ovum
polar body
proliferation
second
sperm
spermatids
spermatocyte
spermatogenesis
zygote

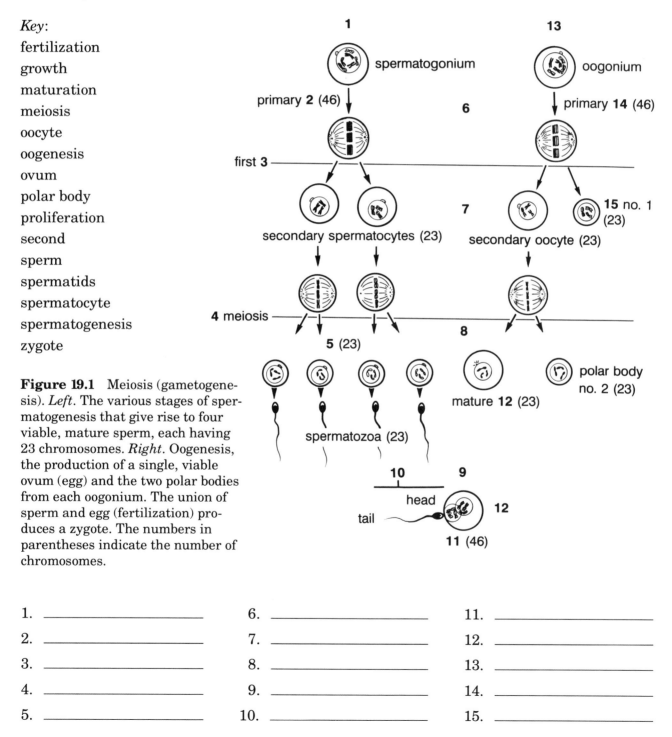

Figure 19.1 Meiosis (gametogenesis). *Left.* The various stages of spermatogenesis that give rise to four viable, mature sperm, each having 23 chromosomes. *Right.* Oogenesis, the production of a single, viable ovum (egg) and the two polar bodies from each oogonium. The union of sperm and egg (fertilization) produces a zygote. The numbers in parentheses indicate the number of chromosomes.

1. _____
2. _____
3. _____
4. _____
5. _____

6. _____
7. _____
8. _____
9. _____
10. _____

11. _____
12. _____
13. _____
14. _____
15. _____

Exercise 19.2

Labeling. Write the name of the structure in the space provided. Color the reproductive tract differently from the surrounding tissue.

Key:

anus	Cowper's	prepuce	scrotum	testis	urethral
bladder	epididymis	prostate	seminal	ureter	vas
corpus	penis	rectum	sigmoid	urethra	

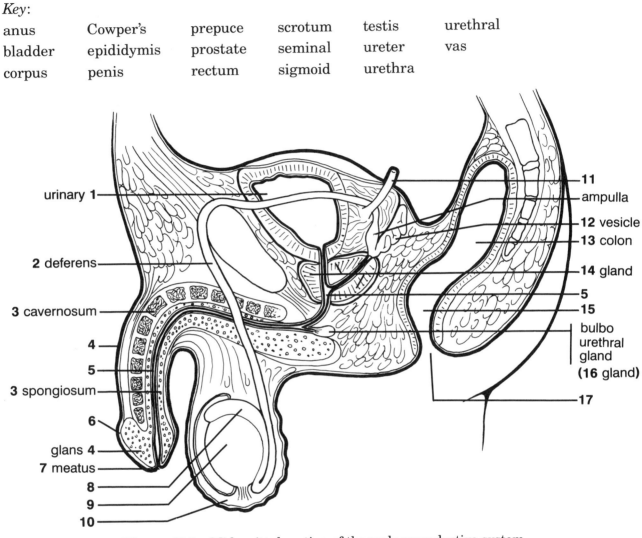

Figure 19.2 Midsagittal section of the male reproductive system.

1. _____   10. _____
2. _____   11. _____
3. _____   12. _____
4. _____   13. _____
5. _____   14. _____
6. _____   15. _____
7. _____   16. _____
8. _____   17. _____
9. _____

Exercise 19.3

Labeling.   Write the name of the structure in the space provided. Color the systems differently.

Key:
anus
bladder
broad
cervix
clitoris
endometrium
Fallopian
hymen
ligament
majora
myometrium
ovary
oviduct
pubis
rectum
sigmoid
urethra
uterus
vagina

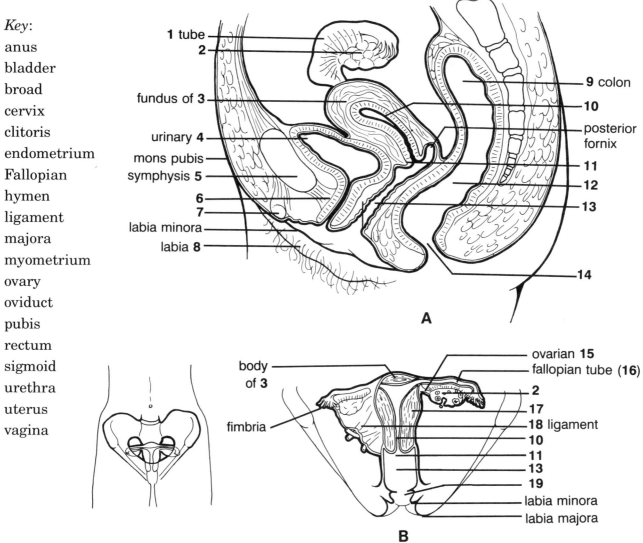

Figure 19.3   Female reproductive system. **A**. Midsagittal section of the reproductive organs of the human female. **B**. Anterior aspect of the female reproductive organs, with the left tube, ovary, and entire ureters cut away to show their internal anatomy. At the left is the position of the female reproductive organs in relation to the pelvis.

1. _____   8. _____   15. _____

2. _____   9. _____   16. _____

3. _____   10. _____   17. _____

4. _____   11. _____   18. _____

5. _____   12. _____   19. _____

6. _____   13. _____

7. _____   14. _____

Exercise 19.4

Labeling.   Write the name of the structure in the space provided. Color the tissues differently.

Key:

afferent

alveolus

ampulla

areola

clavicle

fat

hypothalamus

lactiferous

lobe

muscles

nerves

nipple

pituitary

posterior

ribs

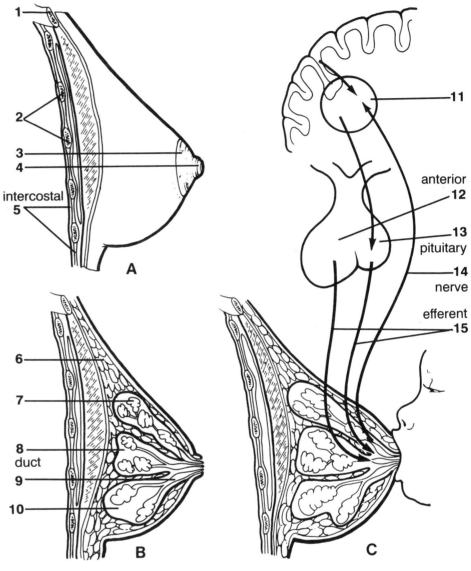

Figure 19.4   Mammary glands. **A**. and **B**. Lateral aspect and sagittal section of the mammary glands, showing the external and internal anatomy of the milk (lactiferous) glands and duct system. **C**. Lactation—The child's suckling gives rise to afferent nerve impulses which, in turn, stimulate the mammary glands to secrete milk.

1. _____

2. _____

3. _____

4. _____

5. _____

6. _____

7. _____

8. _____

9. _____

10. _____

11. _____

12. _____

13. _____

14. _____

15. _____

V. TEST ITEMS

A. *Multiple Choice.*   There is only one answer that is either correct or most appropriate. Circle the answer that corresponds to the question.

1. The male urethra is encircled by which structure?
   a. epididymis
   b. scrotum
   c. prostate gland
   d. seminal vesicle

2. The basic difference between spermatogenesis and oogenesis is that
   a. during spermatogenesis two more polar bodies are produced.
   b. the mature ovum contains the haploid chromosome number, whereas the mature sperm contains the diploid number.
   c. in oogenesis, one mature ovum is produced, and in spermatogenesis, four mature sperm are produced.
   d. spermatogenesis involves mitosis and meiosis, but oogenesis involves meiosis only.

3. Fertilization normally occurs in the
   a. uterine tubes.
   b. vagina.
   c. uterus.
   d. ovaries.

4. The union of a sperm nucleus and an ovum nucleus resulting in the formation of a zygote is referred to as
   a. implantation.
   b. fertilization.
   c. gestation.
   d. parturition.

5. After ovulation, the corpus luteum forms. If the ovum is not fertilized the corpus luteum degenerates but if the ovum is fertilized, the corpus luteum persists for several months. Which hormones help maintain the corpus luteum during the early months of pregnancy?
   a. FSH
   b. ACTH
   c. vasopressin
   d. chorionic gonadotropin

6. What hormone suppresses ovulation during pregnancy?
   a. LH
   b. progesterone
   c. estrogen
   d. FSH

7. The male sex hormone is produced in the
   a. interstitial cells of the testes.
   b. tubules of the testes.
   c. anterior lobe of the pituitary gland.
   d. sustentacular cells of the testes.

8. After ovulation, the ruptured follicle
   a. disappears, and all its cells disintegrate.
   b. passes as waste material down the oviduct with the egg.
   c. mends itself and begins the maturation of another egg.
   d. differentiates into another temporary endocrine gland.

9. The cells lying between sperm-forming cells produce a hormone called
   a. esterone.
   b. testosterone.
   c. progesterone.
   d. aldosterone.

10. Androgens in men are produced by the
    a. prostate.
    b. seminal vesicles.
    c. interstitial tissue in testes.
    d. pituitary.

11. In humans, sperm cells are produced in the
    a. interstitial tissue.
    b. urethra.
    c. seminiferous tubules.
    d. ductus deferens.

12. Which of these is mismatched?
    a. ovary—testes
    b. oviduct—ductus deferens
    c. uterus—epididymis
    d. vagina—penis

13. Semen
    a. contains many sperm in a fluid medium.
    b. is ejaculated during copulation.
    c. is used for artificial insemination.
    d. all of the above

14. The major portion of the volume of semen is contributed by the
    a. bulbourethral glands.
    b. testes.
    c. prostate gland.
    d. seminal vesicles.

15. The chief ligament supporting the position of the uterus and keeping it from dropping into the vagina is the
    a. cardinal ligament.   c. broad ligament.
    b. round ligament.   d. ovarian ligament.

16. The normal site of fertilization is
    a. one-third the way down the uterine tube.
    b. the wall at the fundus of the uterus.
    c. the cervix of the uterus.
    d. the abdominal cavity.

17. The sex chromosomes of a normal male are designated as
    a. YY.   c. XX.
    b. XY.   d. none of the above

18. Sperm
    a. formation results from both meiotic and mitotic division of male germinal epithelium.
    b. carrying an X chromosome will produce genetic males upon fertilization of the female ovum.
    c. are inactivated by an alkaline environment.
    d. removed from the seminiferous tubules are capable of fertilizing an egg.

19. Endometriosis is a disease in which
    a. the lining of the uterus is inflamed.
    b. too few sperm are formed.
    c. endometrial tissue is found in abnormal places.
    d. secretions of the seminal vesicles are too acidic.

20. Estrogen is formed in which of the following?
    a. Graafian follicle   c. corpus luteum
    b. corpus albicans   d. seminal vesicles

B. *Matching Questions.*   Each of the phrases in COLUMN B refers to a word or phrase in COLUMN A. Insert the letter of the word or phrase from COLUMN B that best describes it. Some words may be used more than once or not at all.

Column A		*Column B*
1. ___ mutation	**a.**	removal of the uterine mucous lining
2. ___ abortion	**b.**	a gene that results in embryonic death or death shortly after birth
3. ___ lethal gene		
4. ___ autosome	**c.**	a permanent heritable change in a gene that causes the gene to express a different trait
5. ___ endometrectomy		
6. ___ cesarean section	**d.**	any chromosome that is not a sex chromosome
	e.	removal of a fetus and placenta through an abdominal incision in the uterine wall
	f.	premature expulsion from the uterus of the products of contraception—an embryo or a nonviable fetus

Column A		*Column B*
1. ___ Leydig cells	**a.**	suspended in the scrotum
2. ___ seminal vesicles	**b.**	contains one or more seminiferous tubules
3. ___ spermatids	**c.**	interstitial cells
4. ___ interstitial cells	**d.**	undifferentiated germ cells
5. ___ impotence	**e.**	transformed into mature spermatozoa
6. ___ testes	**f.**	the tip of the sperm
7. ___ ovum	**g.**	a supporting cell type in the testes
8. ___ acrosome	**h.**	drains into the vas deferens
9. ___ antrum	**i.**	corpora cavernosa
10. ___ castration	**j.**	due to psychological disturbance
11. ___ Sertoli cell	**k.**	cells that secrete testosterone
12. ___ lobule	**l.**	removal of the testes
13. ___ corpus luteum	**m.**	the female germ cell
14. ___ erectile tissue	**n.**	a fluid-filled space
15. ___ spermatogonia	**o.**	follicular cells become the

C. *True-False.*   Place a *T* or *F* in the space provided.

____ **1.** In older males, an enlarged prostate gland may obstruct the flow of urine.

____ **2.** The inability of a male to attain or hold an erection is known as infertility.

____ **3.** Low levels of progesterone may cause painful menstruation, a condition known as amenorrhea.

____ **4.** Each duct of a seminal vesicle joins a ductus deferens to form the ejaculatory ducts that open into the prostatic portion of the urethra.

____ **5.** In males, FSH stimulates spermatogenesis and ISCH stimulates the production and secretion of testosterone.

____ **6.** The greater vestibular (Bartholin's) glands lie on either side of the female urethral orifice.

____ **7.** Estrogens are responsible for the preovulatory changes in the uterus following menstruation, and progesterone is responsible for the postovulatory (secretory) phase.

____ **8.** Ovulation is initiated by a sharp rise in LH secretion.

____ **9.** Estrogens are secreted by the ovarian follicle, and estrogens and progesterone are secreted by the corpus luteum.

____ **10.** Spermatogenesis takes place in the epididymis.

____ **11.** Progesterone and estrogens are secreted by the placenta after several weeks of pregnancy.

____ **12.** Estrogens promote the growth of the alveoli of the mammary gland, and progesterone promotes the growth of the ducts.

____ **13.** The disintegration of the corpus luteum (about 10 days after ovulation) is prevented by chorionic gonadotropin.

____ **14.** Estrogens decrease the motility of the uterus and its sensitivity to oxytocin, while progesterone has the opposite effect.

____ **15.** Destruction of the hypothalamus prevents ovulation.

____ **16.** The hypothalamus produces a hormone that travels in the blood to the anterior pituitary, causing the release of luteinizing hormone.

____ **17.** During the process of meiosis, the 46 human chromosomes are separated into 23 male chromosomes and 23 female chromosomes.

____ **18.** Removal of the gonadal tissue of a male embryo prior to the sixth week of development results in the embryo developing female sexual characteristics.

____ **19.** The undifferentiated sex cells (e.g., spermatogonia) increase in number by mitosis during childhood and throughout life.

____ **20.** Mature sex cells are called gametes.

Answer Sheet—Chapter 19

Exercise 19.1

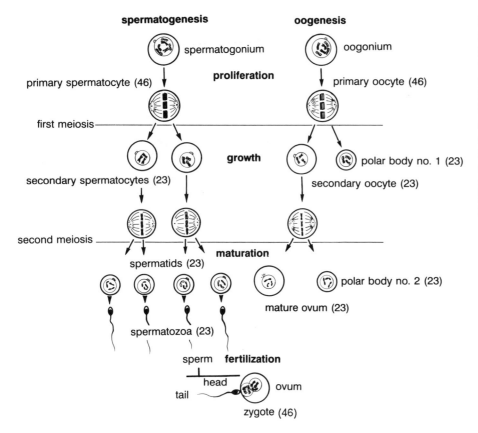

Figure 19.1 Meiosis (gametogenesis). *Left.* The various stages of spermatogenesis that give rise to four viable, mature sperm, each having 23 chromosomes. *Right.* Oogenesis, the production of a single, viable ovum (egg) and the two polar bodies from each oogonium. The union of sperm and egg (fertilization) produces a zygote. The numbers in parentheses indicate the number of chromosomes.

Exercise 19.2

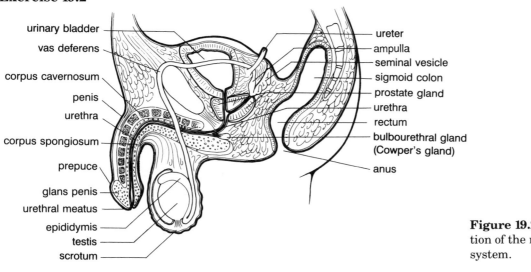

Figure 19.2 Midsagittal section of the male reproductive system.

Exercise 19.3

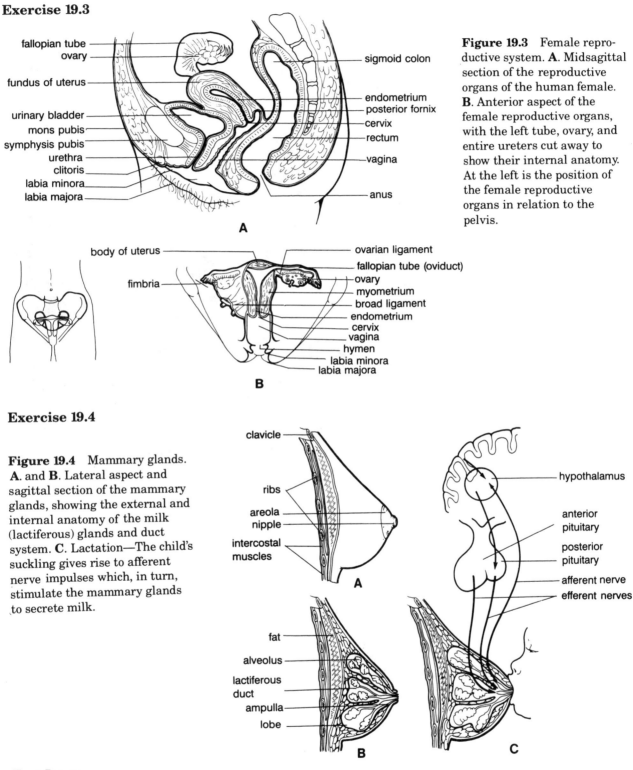

Figure 19.3 Female reproductive system. **A**. Midsagittal section of the reproductive organs of the human female. **B**. Anterior aspect of the female reproductive organs, with the left tube, ovary, and entire ureters cut away to show their internal anatomy. At the left is the position of the female reproductive organs in relation to the pelvis.

Exercise 19.4

Figure 19.4 Mammary glands. **A**. and **B**. Lateral aspect and sagittal section of the mammary glands, showing the external and internal anatomy of the milk (lactiferous) glands and duct system. **C**. Lactation—The child's suckling gives rise to afferent nerve impulses which, in turn, stimulate the mammary glands to secrete milk.

Test Items

A. 1.c, 2.c, 3.a, 4.b, 5.d, 6.b, 7.a, 8.d, 9.b, 10.c, 11.c, 12.c, 13.d, 14.d, 15.c, 16.a, 17.b, 18.a, 19.c, 20.a.

B. 1.c, 2.f, 3.b, 4.d, 5.a, 6.e.
1.c, 2.h, 3.e, 4.k, 5.j, 6.a, 7.m, 8.f, 9.n, 10.l, 11.g, 12.b, 13.o, 14.i, 15.d.

C. 1.T, 2.F, 3.F, 4.T, 5.T, 6.F, 7.T, 8.T, 9.T, 10.F, 11.T, 12.F, 13.T, 14.F, 15.T, 16.T, 17.F, 18.T, 19.T, 20.T.

Reproduction

Across

1 to expel

3 cessation of menstruation

5 union of sperm and egg

8 after the second week of pregnancy

11 chromosome without sexual traits

13 pertaining to body cells

14 sexual haploid cell produced by meiosis

15 sexual intercourse

16 a fertilizable female gamete

17 monthly discharge from the uterus

19 gene containing structure in the nucleus of a cell

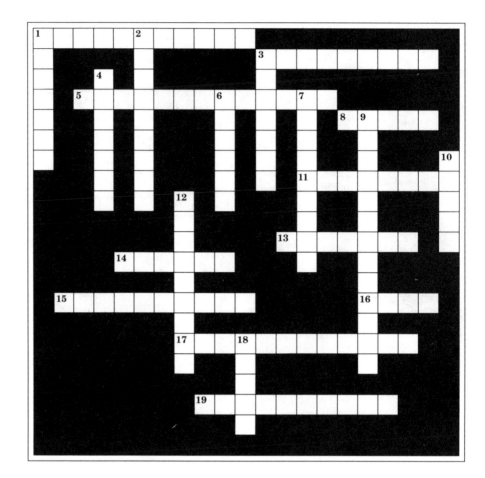

Down

1 out of place

2 secretion of milk from the mammary glands

3 cell division resulting in exact duplication of body cells

4 cell division resulting in haploid cells

6 organism produced by the union of two gametes

7 explusion of an egg from the ovary

9 growth of endometrial tissue outside of the uterus

10 fluid that supports sperm cells

12 surgical excision of the vas deferens

18 male sex gamete

Glossary

abduction movement away from the body midline.

abortion termination of a pregnancy before fetal life is reached.

absorbed dose amount of ionizing radiation absorbed per unit of mass of irradiated material as it passes through matter.

accommodation ability of the eye to adjust the curvature or refractive power of the lens and focus on objects at different distances.

acetylcholine neuronal transmitter substance.

acid any substance containing hydrogen that tends to increase the concentration of hydrogen ions (H^+) in a water solution.

acidosis disorder of body chemistry in which the hydrogen ion concentration of the blood is increased, thus decreasing the pH.

acromegaly disease of the pituitary gland, marked by progressive enlargement of face, hands, and feet—overproduction by the anterior pituitary gland after normal growth has ceased.

actin muscle protein found on the I-band.

active transport movement of a substance across a membrane against the concentration gradient by the expenditure of energy.

adaptation adjustment to changes in the environment.

adduction movement toward the body midline.

adenohypophysis anterior pituitary gland.

adenoma benign glandular tumor.

adenosine triphosphate (ATP) compound composed of one molecule each of adenine and D-ribose with three molecules of phosphoric acid that is concerned with energy transformations in metabolism.

adrenal glands glands located near the kidney that produce hormones.

adrenergic nerve fibers that release a chemical that stimulates the sympathetic nervous system.

adrenocorticotropic hormone (ACTH) a hormone produced by the anterior pituitary that stimulates the adrenal cortex to function.

aerobic reacting chemically or growing only in the presence of oxygen.

afferent toward an organ or area.

agglutination clumping of cells.

agglutinin blood antibody that causes agglutination.

agglutinogen antigen that causes the formation of an antibody.

aging measurement of the degree of physical maturity of a cell or individual.

agranulocyte cell with a clear cytoplasm and no granules.

aldosterone a hormone produced by the adrenal cortex that controls sodium reabsorption by the kidney.

alkalosis disorder of body chemistry in which the hydroxyl ion concentration of the blood is increased, thus elevating the pH.

allergen substance that causes sensitivity.

allergy hypersensitivity to normally harmless substances.

alveolus air cells in bronchi.

amphiarthroses semimovable or partially moving joints.

ampulla saccular dilation of canal.

anabolism constructive phase of metabolism during which protoplasm and other complex compounds such as hormones are synthesized from simpler substances within the cell.

anaerobic chemical reaction or growth without benefit of oxygen.

androgens hormones that stimulate development of male sex characteristics.

anemia lack of blood or lack of a sufficient quantity of red blood cells.

anesthetic drug that produces loss of feeling or sensation.

aneurysm saclike enlargement of a blood vessel caused by a weakening of the vessel wall.

angiography x-ray of vessels after injection with an opaque dye.

ankylosis fixation of a joint.

anorexia loss of apetite.

antagonist muscle that balances the effects of another muscle.

antibody specific substance formed in response to antigen that provides protection against the antigen.

anticoagulant substance that prevents clotting of the blood.

antidiuretic hormone (ADH) hormone secreted by the pituitary gland that controls water absorption in the kidneys.

antigen substance that stimulates the production of antibodies or reacts with them.

aorta the main trunk of the arterial system, emerging from the left ventricle.

apex pointed portion of a conical structure, as in the apex of the lung or heart.

apnea temporary cessation of respiration.

aponeuroses sheets of connective or membranous tissue connecting muscle and the part it moves.

apoplexy sudden loss of consciousness, followed by paralysis due to cerebral hemorrhage.

appendicular skeleton that part of the total skeleton that is suspended from the central supporting frame.

arteriography x-ray of an artery after injection of a radiopaque dye.

arteriole small artery.

arteriosclerosis thickening, hardening, and loss of elasticity of arteries.

artery vessel that carries blood away from the heart.

arthritis inflammation of a joint.

arthrology the science of joints.

articulate to join together so as to permit motion between parts.

articulation the point of union of any two bones.

asphyxiation loss of conciousness due to inadequate oxygen supply; suffocation.

aspiration removal of fluids or gases by suction; the accidental inhalation of an object into the trachea.

assimilation absorption of food; constructive metabolism.

astigmatism faulty vision due to irregular curvature of cornea or lens.

ataxia loss of the power of muscular coordination.

atheroma growth within an artery.

atherosclerosis development of lipid deposits in arterial walls.

atlas first cervical vertebra that supports the head and, with the second cervical vertebra, forms the axis of rotation of the skull.

atom smallest unit of a chemical element.

atrium receiving chamber of the heart.

atrophy wasting or decrease in the size of a part of the body.

audiometer instrument used to measure range of hearing.

autoimmune disease a condition in which antibodies act against the body's own tissue.

autolysis process by which lysosomes break open within the cell, resulting in the self-digestion of the cell.

autosome chromosomes without sex traits.

axial skeleton bones of the vertebral column, thorax, and skull.

axis second cervical vertebra that, with the first cervical vertebra, forms the axis of rotation.

axon extension of a nerve body that conducts an impulse to another body.

axotemia excess amount of urea in the blood.

B cells lymphocytes that produce antibodies and that are derived from bursa or bone marrow.

base any substance containing a hydroxyl group that increases the concentration of hydroxyl ions (OH^-) in a water solution.

benign not a threat to life; nonmalignant.

bilateral found on both sides of the body or body part.

bile bitter, alkaline, yellow-green fluid secreted by the liver.

bilirubin orange pigment derived from hemoglobin.

binocular having two eyes.

biopsy examination of living tissue removed from an organism.

blastula early spherical stage of development of the embryo.

bolus rounded mass of soft consistency.

bond force holding two atoms or ions together in a molecule.

bradycardia abnormally slow heart rate.

bronchus a main branch of the trachea leading to the bronchioles.

buffer any substance in a fluid that tends to resist the change in pH when acid or alkali is added.

bunion swollen, inflamed bursa of the large toe.

bursa fluid-filled sac or space located in areas where friction may develop between moving parts, such as near joints, under muscles, and over bony projections.

bursitis inflammation or irritation of a bursa sac.

calcification process of hardening caused by deposits of calcium compound.

calcitonin hormone produced by the thyroid gland that regulates calcium metabolism.

calculi stones formed within a body part.

callus hard substance formed between fragments of broken bone.

calorie unit of heat; a calorie (cal) is the amount of heat required to raise 1 gram of water from 15° to 16°C.

canal narrow tubular passage or channel.

cancer malignant neoplasm.

capillary small vessels connecting arterioles and venules.

carbohydrate organic compound containing carbon, hydrogen, and oxygen that is used as the major energy source of the body.

carcinoma malignant tumor of epithelial or glandular tissue.

cardiac pertaining to the heart.

caries tooth decay.

casts molds of kidney tubules consisting of protein and blood cells.

catabolism the tearing-down process of metabolism during which complex substances are converted into simpler compounds within the cells.

cataract loss of transparency of the crystalline lens of the eye or of its capsule.

catecholamine amine compounds such as epinephrine and norepinephrine that stimulate the sympathetic nervous system.

celiac relating to the abdominal cavity.

centromere constricted part of the chromosome to which the spindle fibers attach during mitosis.

cerebrovascular accident (CVA) rupture of a vessel in the brain.

cerumen earwax.

cervical referring to the neck.

chemoreceptor receptors that are sensitive to chemical stimuli.

chemotherapy treatment of disease by means of various chemicals.

cholecyst gallbladder.

cholinergic nerve fibers that release a chemical that inhibits muscle action.

chromatin nucleic acid and protein substance that make up a chromosome.

chromosome gene-containing filamentous structure in a cell nucleus.

chyle milky fluid taken up by the lacteals from the food in the intestine after digestion; it consists of lymph and emulsified fat.

chyme semifluid mass of partly digested food passed from the stomach into the duodenum.

circumduction circular movement of a body part.

cirrhosis pathological change in the connective tissues of an organ, particularly in the liver.

cisterna enclosed space.

cleavage splitting of a molecule or fertilized egg during cell division.

clonus spasm in which a muscle contracts and relaxes alternately.

coagulation clotting.

coarctation out of position.

codon three adjacent nucleotides coding a specific amino acid.

coenzyme nonprotein substance that activates an enzyme.

coitus sexual intercourse.

collagen chief structural protein of skin and connective tissue.

colostrum first fluid secreted from the mother's breast following childbirth.

colposcopy vaginal examination.

coma state of profound unconciousness from which one cannot be aroused.

comedone blackhead.

commissure group of nerve fibers connecting opposite corresponding portions of the central nervous system.

compound substance in which the molecules are composed of two or more different elements.

computerized fluoroscopy computerized analysis of a direct examination of the body using a fluoroscope.

computerized tomography (CT) a process of cross-sectional x-ray examination that utilizes a computer for analysis.

concave curved inward.

concentration gradient parts per volume.

conductance transmission of an impulse from one point to another.

conduction transmission of energy from one point to another.

condyle rounded projection.

cones photoreceptors in the eyes that are responsible for detailed vision and color.

connective tissue supporting tissue of the body.

constipation decreased bowel action.

contractility ability of a muscle to become short and thick.

convex curved outward.

convulsion involuntary contraction or series of contractions of voluntary muscles.

copulation sexual intercourse.

cortex outer portion of an organ.

cortisone hormone that regulates carbohydrate metabolism.

covalent bond sharing of electrons by two atoms in a chemical compound.

creatine phosphate (CP) an intermediate source of energy in muscle contraction.

crenation shriveling of a cell due to the passage of its fluid into the surrounding medium.

crest ridge surrounding a bone or its border.

cryptorchidism failure of the testes to descend into the scrotum.

curie unit of radioactivity.

cutaneous skin.

cyanosis dark, bluish appearance of the skin, lips, and nails due to inadequate oxygenation of the blood.

cytoscope instrument used to examine the inside of the bladder.

deafness impairment or defect in the ability to hear sounds.

deamination removal of an amino group from a substance.

defecation expulsion of fecal matter from the bowel.

defibrillator device to correct ventricular fibrillation.

deglutition swallowing.

dehydration inadequate water content in the body tissues.

dementia deterioration of intellectual function.

dendrite extension of a nerve body that conducts an impulse toward the nerve body.

deoxyribonucleic acid (DNA) type of ribonucleic acid that is considered the basic material of life and is involved in the genetic code.

dermatitis inflammation of the skin.

dermatosis general term for any skin disorder.

dermis connective tissue layer under the epidermis.

dextroposition out of position.

dialysis separation of crystalloids from colloids in solution by using a selectively permeable membrane.

diapedesis passage of blood cells, especially white cells (leukocytes), through the intact blood vessel walls.

diaphysis shaft of a long bone.

diarrhea abnormally liquid discharge from the bowels.

diarthroses joint that permits free movement.

diastole rhythmic period of relaxation and dilation of the heart.

dichromatic having two-color vision instead of three-color vision.

diffusion passage of a liquid or gas from a region of greater concentration of its molecules to a region of lesser concentration of its molecules.

diplopia double vision.

dislocation displacement of a limb or organ from its original position.

diureses urine output.

diverticulum outpouching from a main tubular structure or organ cavity.

dominant trait that will be expressed.

dorsal pertaining to the back; posterior.

dose rate radiation dose delivered per unit of time, usually in roentgens per minute.

dosimeter device that measures radiation exposure (e.g., film badge, ionization chamber, Geiger counter).

dysfunction impaired function.

dysmenorrhea painful menstruation.

dysphagia difficulty in swallowing.

dyspnea difficult or labored breathing.

dystrophy degenerative disease of the body tissues.

dysuria painful urination.

ectopic out of the normal place.

edema swelling caused by accumulation of excess fluid.

efferent away from an organ or area.

ejaculation ejection of semen.

elasticity ability to be stretched and return to normal shape.

electrocardiogram (ECG) graphic record of the electrical activity of the heart muscle.

electrolyte solution containing free ions and therefore having the ability to conduct an electrical current.

electron particle in motion outside the nucleus of an atom that carries a negative charge.

elements substances that cannot be decomposed or transformed by chemical means.

elephantiasis enlargement of tissue due to inadequate tissue drainage.

embolism blocking of a vessel by a clot or foreign material.

eminence prominence or projection.

emphysema dilation of the pulmonary air vesicles, usually through atrophy of the septa between the alveoli.

emulsification mechanical digestion of fats by bile.

endocardium inner membrane of the heart.

endochondral within cartilage.

endocrine glands ductless glands whose secretions are transported via bloodstream.

endometriosis proliferation of endometrial tissue outside of the uterus.

endorphin peptide found in the pituitary and involved in pain inhibition.

endoscope instrument used to look inside hollow organs.

endoskeleton internal supporting bony framework.

endothermic heat- or energy-absorbing.

enkephalin peptide found in the brain and involved in pain inhibition.

enteric pertaining to the intestines.

enzyme organic catalyst that is made in a cell.

epicardium outer membrane of the heart.

epicondyle projection above a condyle.

epidermis outer layer of the skin.

epimysium sheath of connective tissue surrounding individual muscles.

epinephrine hormone that stimulates the sympathetic nervous system.

epiphysis ends of long bones.

epistaxis nosebleed.

epithelium nonvascular cellular layer that covers the internal and external surfaces of the body.

equilibrium state of balance, resulting in a stable system.

erythema redness of the skin.

erythrocyte a red blood cell.

erythropoiesis manufacture of red blood cells.

estrogens hormones that stimulate development of female sex characteristics.

eunuch castrated male.

eupnea normal, easy respiration.

eversion turning a body part away from the body midline.

excitability ability to respond to a stimulus.

excretia discharged natural waste.

excretion separation and removal of substances by the cell.

exocrine glands glands with ducts whose secretions are transported outside the body.

exoskeleton external supporting bony framework.

exothermic heat- or energy-releasing.

expiration movement of air out of the lungs.

extensibility ability of a muscle to be stretched.

extension straightening a limb or body part.

exteroceptors nerve endings that detect environmental changes that directly affect the skin.

extracorporeal outside the body.

extrasystole premature contraction of heart muscle.

facet small flat surface.

familial affecting several members of the same family.

fascia sheet of connective tissue.

fasciculus small bundle of nerve fibers or muscle cells.

fatigue inability to respond to a stimulus.

fats compound made of glycerol and fatty acids; lipid.

fertilization union of the ovum and sperm.

fetus name given to a developing human organism after the second month of pregnancy.

fibrillation uncoordinated contractions of individual muscle fibers.

fibrositis inflammation of connective tissue.

filtrate liquid that has passed through a membrane or filter.

filtration passage of a liquid through a filter or membrane by a force that is exerted on the mixture.

fissure relatively deep cleft or groove.

flaccid flabby, soft.

flare diffuse redness of the skin surrounding an injured or pressured point.

flexion bending a limb or body part.

flutter fast, irregular motion.

follicle-stimulating hormone (FSH) pituitary hormone that stimulates development of the egg in females and sperm in males.

fontanel unossified area between cranial bones.

foramen natural opening or passage.

fossa trench or channel, which denotes a hollow or depressed area.

fracture broken bone.

gamete sexual haploid cell produced by meiosis.

gamma ray electromagnetic radiation that originates from a radioactive nucleus and causes ionization in matter.

ganglion aggregation of nerve cells within the brain along the course of a sensory nerve.

gastrulation formation of the third embryonic germ layer.

gene unit of heredity located in the chromosome and made mostly of DNA.

glands secreting organs.

glaucoma disease characterized by abnormally high pressure within the eye, resulting in blindness.

glucagon hormone that aids in the breakdown of glycogen in the liver.

glucocorticoids steroid hormones that stimulate production of glucose from noncarbohydrate sources.

gluconeogenesis formation of glucose from noncarbohydrate sources.

glycocalyx carbohydrate-rich outer covering on the surface of cells.

glycogenesis formation of glycogen from simple sugars.

glycogenolysis breakdown of glycogen into simple sugars.

glycoprotein carbohydrate-protein compound; a conjugated protein.

glycosuria presence of glucose in urine.

goiter enlargement of the thyroid gland.

granulocyte cell with granules in the cytoplasm.

gustation sense of taste.

gyrus smooth surface of an organ.

half-life time (specific for each radioactive substance) required for radioactive material to decay to half its initial activity.

helical spiral.

hematocrit formed element content of the blood.

hematopoietic producing blood cells.

hemodialysis removal of wastes from blood through a semipermeable membrane.

hemodynamics forces connected with circulation of blood.

hemoglobin an iron-protein compound that carries gases in the blood.

hemolysis disintegration of red blood cells, which results in the appearance of hemoglobin in the surrounding fluid.

hemophilia sex-linked, hereditary disease characterized by prolonged coagulation time and abnormal bleeding.

hemorrhage bleeding through vessel walls.

hemostasis checking flow of blood through any part of the body.

hermaphroditism condition of having both male and female sex organs.

hernia weakened opening in the abdominal wall.

heterolysis destruction of cells of one species by lysins of a different species.

histamine vasodilating substance in many cells.

histocompatibility tolerance of host tissue to donor or foreign tissue, such as occurs in transplants.

homeostasis consistency and uniformity of the internal body environment, which maintains normal body functions.

hormone chemical substance produced in one organ that, when carried to another organ by the circulation, stimulates the latter organ to functional activity.

hyaline glassy membrane found in the newborn lung.

hydrocortisone steroid hormone that helps the body reduce stress.

hydronephrosis accumulation of urine in the kidney due to an obstruction.

hydrophilic affinity for water.

hydrophobic tending to repel water.

hydrostatic pressure pressure created by fluid content.

hyperactive increased activity; overactive.

hypercapnea high carbon dioxide content of the air or blood.

hyperemia swelling due to increased blood supply.

hyperglycemia excess of sugar in the blood.

hyperkalemia elevated potassium concentration in the blood.

hypermenorrhea prolonged menstruation.

hyperopia farsightedness.

hyperplasia increased size of an organ or tissue.

hyperpnea abnormally rapid respiratory movements.

hypertension elevated blood pressure.

hypertonic having a higher osmotic pressure than some other solutions.

hypertrophy increase in size of a tissue or organ.

hypoactive diminished activity; underactive.

hypogastric positioned below the stomach region.

hypoglycemia deficiency of sugar in the blood.

hypophysis pituitary gland.

hypotonic having a lower osmotic pressure than some other solutions.

hypoxia insufficient oxygen in the body tissues.

immunity properties of the host that protect it from foreign agents.

immunodeficiency disease disease due to failure of some immune function.

immunoglobulin antibody against a particular antigen; protective immunity.

immunosuppression use of drugs to weaken immune response.

incontinence inability to control the passage of urine or feces.

infarction death of tissue due to loss of blood supply.

infectious capable of producing disease in a susceptible host.

inflammation series of reactions in tissues produced by microorganisms or other irritants and marked by redness of the affected area.

insertion attachment of a muscle to the more movable bone.

inspiration active mechanism creating a vacuum in the lungs.

insulin hormone that regulates carbohydrate metabolism.

integument covering, especially the skin.

interoceptors receptors within organs concerned with the maintenance of the internal environment.

interstitial between cells.

interstitial-cell-stimulating hormone (ICSH) pituitary hormone that stimulates androgen production in the testes.

intoxication pathological state produced by a drug, serum, alcohol, or any toxic substance.

intrapulmonary space within the alveolar sacs.

intrathoracic space in the thoracic cavity between the pleura.

intussusception infolding of one segment of the intestine within another segment.

inversion turning a body part toward the body midline.

involuntary performed without free will.

ion charged particle.

ionization production of ions from neutral atoms or compounds.

ischemia lack of blood in an area of the body.

isometric contraction of a muscle without shortening its length.

isotonic condition of equal osmotic pressure between two different solutions.

isotope element that has the same atomic number as another but a different atomic weight.

jaundice yellowness of skin and eyes.

joint point of connection between two or more bones.

keratin tough fibrous protein produced by keratinocytes.

kinesthetic referring to the ability to sense movement.

kyphosis increased curvature of the thoracic spine, giving a hunchback appearance.

lactation secretion of milk by the mammary glands.

lacteal one-celled vessel of the lymphatic system.

lactogenic hormone (LTH) pituitary hormone that stimulates milk production.

lacuna small hollow, depression, or pit.

lamina thin, flat layer in a portion of tissue, consisting of layers of cells; also a flat plate (e.g., the laminae of vertebrae).

leukemia disease of the blood-forming tissues marked by increase in the number of leukocytes (leukocytosis).

leukocyte white blood cell.

leukocytosis increase in the number of leukocytes caused by the host body's response to an injury or infection.

leukopenia decrease in the number of leukocytes.

leukorrhea vaginal discharge other than blood.

ligament band of fibrous tissue that connects bones and strengthens joints.

lipid fat, oil, or their derivatives.

lordosis forward curvature of the lumbar spine.

lumbar referring to the lower back.

luteinizing hormone (LH) pituitary hormone that stimulates formation of the corpus luteum in the ovary.

lymph a fluid found in the tissue spaces that contains most of the components of blood.

lymph nodes oval structures located along a lymphatic vessel that filter foreign matter and produce lymphocytes.

lymphangiogram injection of an opaque dye into a vein for x-ray purposes.

lymphatic pertaining to lymph nodes.

lymphokines soluble substances produced by lymphocytes that can affect other cells.

lymphoma proliferation of lymphatic tissue.

lysis rupture of a cell.

macrocytic large cell.

malignant referring to disorders that tend to worsen and cause death.

malleolus hammer-shaped protuberance.

mastectomy removal of breast tissue.

mastication act of chewing food.

matrix intercellular substance of a tissue.

matter substance.

mediastinum   wall dividing the thoracic cavity.

medullary   centrally located soft tissue.

meiosis   special method of cell division occurring during the development of sex cells (ova and sperm) in which the number of chromosomes is reduced.

melanin   dark pigment found in skin, hair, and retina.

melanocyte-stimulating hormone (MSH)   pituitary hormone that influences production of melanin.

melanocytes   pigment cell of the skin that produces melanin.

menarche   time of life when the menstrual cycle begins.

meninges   three membranes that envelop the brain and the spinal cord.

menopause   period of life when menstruation normally ceases; change of life.

menorrhagia   excessive menstrual flow.

menorrhalgia   painful menstruation.

menstruation   monthly event characterized by a bloody discharge from the uterus.

mesenchyme   embryonic connective tissue.

metabolism   physical and chemical processes by which living organisms produce the necessary energy to maintain life.

metric   relating to the meter as a basis of measurement.

metrorrhagia   irregular bleeding from the uterus.

microcytic   small cell.

micturition   urination.

mineralocorticoids   steroids that control salt metabolism.

mitosis   form of nuclear division characterized by complex chromosome movements and exact chromosome duplication.

mixture   two or more substances that are not chemically combined.

molecule   smallest unit of a compound that exists in nature.

monochromatic   having only one-color vision.

monosomy   when one of a pair of homologous chromosomes is missing.

mosaic   inlaid network of pattern of small pieces.

motor end plate   axonic terminals of motor neurons.

motor unit   that which produces movement.

mucin   substance secreted by mucous membranes that contains mucopolysaccharides.

mucoprotein   compound composed of proteins and mucopolysaccharides.

murmur   abnormal sound indicating a pathological condition of a heart valve.

muscle   fibrous bands connected to bone that produce movement by contracting.

myalgia   muscle pain or aching.

myelin   fatty protective sheath around a nerve.

myelocytic   produced in the bone marrow.

myeloma   tumor found in the bone marrow.

myocardium   part of the heart wall made up of muscle.

myofibril   contractile fibers within a muscle fiber.

myopathy   disease of the muscles.

myopia   nearsightedness.

myosin   muscle protein found on the A-bands.

myositis   inflammation of a muscle.

necrosis   tissue death, usually in a localized area.

neoplasm   new growth; a tumor.

nephron   functional unit of the kidney.

nerve fiber   extension of the nerve body.

neuralgia   severe pain along the course of a nerve.

neuritis   inflammation along a nerve.

neuroglia   supporting cells to the nervous system.

neurohypophysis   posterior pituitary gland.

neuron   basic functional unit of the nervous system.

neurotransmitters   chemical substance able to transmit an impulse between two structures.

neutron   particle found in the nucleus of an atom that is neutral (i.e., does not carry a charge).

norepinephrine   hormone that stimulates the sympathetic nervous system.

nucleic acid   one of a class of molecules composed of joined nucleotide complexes; the principal types are deoxyribonucleic acid (DNA) and ribonucleic acid (RNA).

nutrition   utilization of food for growth.

nystagmus   involuntary side-to-side movements of the eyes.

obesity   excess fat.

occlusion   closing of an opening or passage.

oncology   study and treatment of tumors.

oncotic pressure   osmotic pressure exerted by colloids.

oocyte   immature ovum or egg cell.

oogenesis   process of formation of ova or egg cells.

ophthalmoscope   instrument used to visualize the retina.

opsonization combination of antibody and antigen that makes them susceptible to phagocytosis.

organelle tiny specific particle of living material present in most cells and serving a specific function in the cell.

orgasm culmination or climax of sexual intercourse.

origin attachment of a muscle to the less movable bone.

osmosis passage of molecules of a pure solvent, such as water, from a solution of lesser concentration to one of greater concentration.

osseous bony or bonelike.

ossification process of forming bone or the conversion of fibrous tissue or cartilage into bone.

osteoblast young bone-forming cell.

osteoclast cell that absorbs bone tissue.

osteocyte mature bone cell.

osteogenic derived from bone.

otoliths calcium carbonate masses of the inner ear.

ovulation expulsion of the ovum from a follicle in the ovary.

ovum female sex gamete.

oxytocin hormone from the posterior pituitary that stimulates smooth muscle contraction.

palsy loss or impairment of nerve or muscle function.

pancreas abdominal gland that secretes enzymes for digestion and hormones that regulate carbohydrate metabolism.

papilla any small projection or elevation.

paralysis loss of muscle function; inability to move.

parathyroid glands a set of small glands behind the thyroid gland that produces a hormone to regulate calcium level in the blood.

particles small portions of matter.

particulate radiation pertaining to having small particles that emit energy.

parturition birth.

pathogenic capable of producing disease.

peduncle group of nerves.

pepsin enzyme produced in stomach responsible for chemical breakdown of proteins.

perfusion passage of fluid through the vessels of an organ.

pericardium serous membrane that lines the sac enclosing the heart and attaches to the heart itself.

perichondrium fibrous membrane that covers cartilage.

periosteum fibrous membrane that covers bone tissue.

peristalsis rhythmic waves of smooth muscle contractions.

peritoneum large serous membrane that lines the abdominal cavity and covers the organs within the abdominal cavity.

permeability the extent to which molecules of various kinds can pass through cellular membranes.

permeable membrane that allows passage of all particles.

phagocytosis process by which a cell engulfs and digests a particle or substance.

phosphocreatine source of energy found in muscle.

photon unit of energy of a light wave.

photoreceptor receptor sensitive to visible light.

pigmentation coloration by deposition of pigments.

pineal gland cone-shaped gland in the middle of the brain that produces melatonin, which inhibits secretions of male sex hormones.

pinocytosis process by which a cell engulfs and digests a droplet of liquid.

pituitary gland almond-shaped gland at the base of the brain that produces hormones that regulate many body functions.

placebo chemical substance given in place of medication.

placenta organ within the uterus through which the fetus derives its nourishment.

plasmalemma flexible cell membrane.

platelet a thrombocyte.

pleura membrane covering the lungs.

pleurisy inflammation of the pleura.

plexus network or tangle of interweaving nerves, veins, or lymphatic vessels.

pneumonia inflammation of the lungs.

pneumothorax air in the thorax.

polycythemia abnormally large number of red blood cells.

polymerize process of joining small compounds to form a compound of high molecular weight.

polyuria abnormally large quantity of urine.

precipitation conversion of a soluble substance to an insoluble substance.

presbyopia vision of older adults.

pressoreceptors receptors sensitive to mechanical stimuli.

process prominence or projection.

progestin hormone of the ovary that stimulates uterine (endometrial) development.

prolactin hormone of the pituitary gland that stimulates milk secretion.

pronation lying face down or moving the arm so that the palm of the hand is facing backward.

proprioceptors receptors that provide the body with information about its position in space.

prostaglandin hormonelike chemical with a variety of effects.

protein complex nitrogenous compound of high molecular weight.

proton particle found in the nucleus of an atom that carries a positive charge.

protoplasm building material of all organisms.

protraction movement of the jaw forward.

pruritis itching.

pseudostratified a false layered effect.

pulse rhythmic contraction and relaxation of muscles along a vessel.

pustule small, pus-containing elevation on the skin.

pyogenic producing pus.

quantum unit of light energy.

radiation emission and projection of energy.

radiation therapy medical treatment with ionizing radiations.

radical two or more atoms reacting as one in a chemical compound.

radioactive pertaining to atoms of elements that undergo spontaneous transformation, resulting in emission of radiation.

radiography photographic film produced by x-ray.

radioisotope radioactive isotope of an element.

radioresistance resistance of cells to radiation.

radiosensitivity responsiveness of cells to radiation.

ramus branch.

receptor sensory nerve ending that responds to stimuli.

recessive trait that is not expressed.

refraction bending of light rays.

releasing factors hormonelike chemicals that stimulate a gland to release its hormone into the bloodstream.

respiration a physical and chemical process in which the organism takes in oxygen, uses it to produce energy, and releases a waste product—carbon dioxide.

response reaction to a stimulus.

resuscitation restoration to consciousness after respiration has ceased.

reticulum a fine network of material.

retraction movement of the jaw backward.

retroperitoneal located behind the peritoneum.

rhodopsin photoreceptor chemical of the rods in the eye.

ribonucleic acid (RNA) type of nucleic acid that plays a role in synthetic reactions within the cell.

rods photoreceptors in the eyes involved with the detection of faint light.

rotation movement around a vertical axis.

sarcolemma cell membrane of a muscle cell.

sarcoma malignant tumor of connective tissue.

sarcomere structural and functional unit of a myofibril.

sarcoplasm protoplasmic material found in a muscle fiber.

scan record of radioactivity within an organ.

scoliosis lateral curvature of the spine.

sebum secretion of a sebaceous gland.

seizure attack, as of a disease or of convulsions.

semen ejaculatory fluid consisting of sperm cells and secretions of the prostrate and bulbourethral glands and the seminal vesicles.

semipermeable permitting the passage of some particles (molecules) and not others.

sense organ any organ of special sense.

septum partition.

sesamoid seedlike bone.

sexome sex chromosome.

shock acute peripheral failure of blood circulation.

sinus cavity or hollow space.

skeletal muscle a striated, voluntary muscle that is attached to bones.

smooth muscle unstriated, involuntary muscle found in organs and blood vessels.

solute substance to be dissolved in a solution.

solvent solution or vehicle capable of dissolving a substance.

somatic pertaining to body cells.

somesthetic bodily awareness.

spasm involuntary, convulsive muscular contraction.

sperm male sex gamete.

spermatogenesis process of formation and development of the spermatozoa.

sphygmomanometer instrument for measuring arterial blood pressure.

spine thornlike process or projection.

splenomegaly enlarged spleen.

sprain joint injury resulting from wrenching or twisting.

squamous scalelike.

stenosis narrowing of an opening, duct, or canal.

steroid chemical substance made up of four interlocking rings of six-carbon atoms.

sterol unsaturated alcohol.

stimulus irritant or excitant.

strabismus squinting.

stratified layers of tissue.

stretch receptors sensory receptors that recognize mechanical distension.

stroke blockage of a cerebral blood vessel.

subcutaneous located beneath the skin.

substrate any substance acted upon by an enzyme.

sulcus depression or separation.

supination lying face up or moving the arm so that the palm of the hand is facing forward.

suppuration formation of pus.

suture junction line between two immovable bones.

symphysis partially movable joint between two bones.

synapse region where parts of two neurons are anatomically related so that impulses are transmitted from one neuron to another.

synarthroses immovable joint between two bones.

syncytium mass of cytoplasm with several nuclei.

synergist two or more structures working together.

synovia viscid fluid of a joint or similar cavity.

synovial of or pertaining to synovia.

systole contraction of the heart muscle.

T cells lymphocytes originating from the thymus.

tachycardia excessively fast heart rate.

taste buds receptors for taste on the tongue.

teleceptors nerve endings that detect environmental changes occurring some distance from the body.

tendon tissue that connects muscle to bone.

tenosynovitis inflammation of the sheath around a tendon.

testosterone a hormone responsible for growth and development of the male sex organs.

tetany intermittent tonic muscular contractions of the extremities.

thermodynamics concerned with heat and its conversion to other forms of energy.

thoracic referring to the thorax.

thorax the chest.

thrombocytopenia lack of platelets, causing hemorrhages.

thrombophlebitis condition in which inflammation of the vein wall has preceded the formation of a thrombus or intravascular clot.

thrombus blood clot.

thymus gland a lymphoid structure at the top of the sternum responsible for immunity in children.

thyroid gland gland found in the neck that produces a hormone that controls the basal metabolic rate.

thyroid-stimulating hormone (TSH) hormone from the brain that controls the release of thyroid hormones.

thyrotropin hormone from the pituitary that stimulates the growth and function of the thyroid.

thyroxin iodine-containing hormone from the thyroid that regulates body metabolism.

tinnitus ringing or singing sound in the ears.

tissues groups of cells similar in origin, structure, and function.

tonus partial, continual contraction of a muscle.

toxins poisonous substances produced by pathogenic organisms.

trabeculae supporting network of tissue fibers.

transamination addition of an amino group to a substance.

transcription transfer of information from the chromosome to the ribosome.

translation process of directing the production of proteins using mRNA at the ribosome.

trauma an injury or wound that may be produced by external force or shock.

trichromatic having normal three-color vision.

triiodothyronine (T3) thyroid hormone that aids in maintaining body metabolism.

tropomyosin muscle protein that inhibits contraction.

troponin muscle protein that inhibits contraction.

tubercle nodule or small, rough, rounded eminence.

tuberosity elevation or protuberance.

tumor new growth or neoplasm.

ulcer lesion of a mucous membrane.

ultrafiltration filtration capable of removing all particles, except viruses.

urea waste product of protein metabolism.

uremia presence of toxic substances in the blood.

urination act of voiding urine.

valence   combining power of an atom.

varices   abnormally swollen.

varicose veins   unnaturally swollen and tortuous veins.

vascular   pertaining to or consisting of vessels.

vasectomy   surgical excision of the vas deferens.

vasoconstriction   narrowing of blood vessels by a nerve or a chemical substance.

vasodilation   widening of the blood vessels.

vasomotor   regulating the contraction and dilation of the blood vessels.

vasopressin   hormone from the posterior pituitary that prevents major water loss by the kidney.

vein   vessel that carries blood to the heart.

ventilation   process of bringing air into and out of the lungs.

ventral   pertaining to the front; anterior.

ventricle   pumping chamber of the heart.

venule   small veins.

vertebra   one of 33 bones that form the spinal column.

vertigo   dizziness.

vesicle   small sac or blister filled with fluid.

vital capacity   volume of air that can be expelled following full inspiration.

vitamin   organic substance necessary for normal metabolism.

wart   small, horny outgrowth of the skin.

wheal   ridgelike swelling of the skin.

zygote   organism produced by the union of two gametes.